T0258879

ANALYTICAL FRAMEWORK FOR INTEGRATED WATER RESOURCES
MANAGEMENT

IHE MONOGRAPH 2

Analytical Framework for Integrated Water Resources Management

Guidelines for assessment of institutional frameworks

PAUL J.M. VAN HOFWEGEN & FRANK G.W. JASPERS
International Institute for Infrastructural,
Hydraulic and Environmental Engineering, Delft

A.A.BALKEMA/ROTTERDAM/BROOKFIELD/1999

Published by
A.A. Balkema, P.O. Box 1675, 3000 BR Rotterdam, Netherlands
Fax: +31.10.4135947; E-mail: balkema@balkema.nl; Internet site: http://www.balkema.nl

A.A. Balkema Publishers, Old Post Road, Brookfield, VT 05036-9704, USA
Fax: 802.276.3837; E-mail: info@ashgate.com

ISBN 90 5410 472 4 hardbound edition
ISBN 90 5410 473 2 student paper edition

Table of contents

PREFACE   VII

ACKNOWLEDGEMENTS   VIII

ABBREVIATIONS AND ACRONYMS   IX

SUMMARY   XI

1   THE ANALYTICAL FRAMEWORK
     1.1   Introduction   1
     1.2   Functions and Functional Levels in IWRM   1
     1.3   The Analytical Framework   2
     1.4   Methodological Steps   3

2   INTEGRATED WATER RESOURCES MANAGEMENT - GENERAL
     PRINCIPLES
     2.1   Aim and Definition   7
     2.2   Water Systems   8
     2.3   Need for Integrated Water resources Management   8
     2.4   Sustainable Use of Water Resources   9
     2.5   Interests   9
     2.6   The "Ideal" IWRM Situation   11

3   IWRM REQUIREMENTS FOR THE FUNCTIONAL LEVELS
     3.1   The Constitutional Function: Water Policy and Law   13
     3.2   The Organizational Function: Integrated Water Resources Management   21
     3.3   The Operational Function: Water Services   25

4   INTERVENTIONS or IWRM CAPACITY BUILDING
     4.1   The Concept of Capacity Building   33
     4.2   The Process   34
     4.3   Models for Capacity Building   34
     4.4   Instruments for Capacity Building and Human Resources Development   35
     4.5   Human Resources Development - Education and Training 35

5   ASSESSMENT OF THE INSTITUTIONAL FRAMEWORK - PROCESS AND
     TOOLS
     5.1   Existing Water Management Situation   39
     5.2   Stakeholder Selection   43
     5.3   Stakeholder Interviews   43

5.4   Analysis Stakeholder Opinions   45
5.5   Workshop 1 - Problem identification   45
5.6   Workshop 2 - Formulation of Desired IWRM Situation and Interventions   46
5.7   Preliminary Country Report   47
5.8   Dissemination and Comments   47
5.9   Final Country Report   47
5.10   Monitoring Procedure.   47

6   APPLICATION OF THE GUIDELINES AND LESSONS LEARNED   49

REFERENCES

ANNEXES

1   GUIDELINES FOR ASSESSMENT OF INSTITUTIONAL FRAMEWORK -   55
        The Constitutional Function

2   GUIDELINES FOR ASSESSMENT OF INSTITUTIONAL FRAMEWORK -   67
        The Organizational Function

3   GUIDELINES FOR ASSESSMENT OF INSTITUTIONAL FRAMEWORK -   79
        The Operational Function

4   EVALUATION MATRIX   93

VI

Preface

The Inter-American Development Bank (IDB) developed a strategy for Integrated Water Resources Management (IWRM) which aims to help borrowing member countries to shift from a sectoral, development-based focus to an integrated, management-based approach. One key principle of this strategy is an increased emphasis on institutional issues and capacity building.

The Bank contracted the International Institute of Infrastructural, Hydraulic and Environmental Engineering (IHE) in Delft, the Netherlands (IDB-ITN/NE-5661-RG) to carry out a consultancy to develop an analytical framework and guidelines for the assessment of an institutional framework for integrated water resources management. This framework was to be tested in four member countries (Guatemala, Jamaica, Colombia and Chile). The information regarding Chile is compiled through a desk study.

The objective of the technical cooperation was to develop an analytical framework for the assessment of the institutional setting for integrated water resources management. The Bank could use this for the incorporation of capacity building considerations in future IDB water related projects. The framework and the guidelines are made for project teams, bank officers and government agencies to facilitate the process of project formulation and monitoring. It also should improve the integration of different steps in the project cycle of project managers and bank field offices and generate more their involvement.

Integrated Water Resources Management (IWRM) means decision making on development and management of water resources for various uses taking into account the needs and desires of different users and stakeholders. This means that the process to identify interventions for institutional development and capacity building for IWRM must be consultative and participatory. This was the basis of a preliminary analytical framework with general guidelines made to be used by the local consultants for their first inventories in the different countries. During the inventory the practical use was tested and gaps and shortcomings were identified. A corrected 1st draft of the analytical framework and guidelines was prepared to be presented and improved in a series of local workshops in which the different stakeholders from the respective countries participated. The objective of these workshops was to test the draft analytical framework and guidelines through the country reports. The outcome of these workshops has been analysed and incorporated in the 2nd draft of the general framework and guidelines. The process was concluded with a workshop at IDB Headquarters in Washington in which Bank staff participated. The outcome is incorporated in this document.

This document gives an analytical framework and general guidelines to be used by IDB-teams in the assessment of the institutional framework in the different IDB member countries and the identification of interventions directing towards integrated water resources management. It presents an integration of two concepts that are in many countries new or under development: Integrated Water Resources Management (IWRM) and Capacity Building. In chapter 1 the analytical framework is presented in which these two concepts are embedded. Another element on which the analytical framework is based is the IDB strategy on Integrated Water Resources Management. For proper assessment of the required capacity building interventions for IWRM, the general principles of IWRM should be well understood. Chapter 1 provides a general outline of these principles. In chapter 2 the institutional requirements are explained for IWRM at the three functional levels (constitutional, organizational and operational). Based on these requirements the concept of capacity building and possible interventions for these three levels are presented in chapter 3. The process to come to the assessment of the present institutional framework and the capacity building interventions is described in Chapter 4. This process consists of 10 steps to be carried out by an IDB expert team. In chapter 5 the lessons learned from the four case studies are presented with points that should be considered before application of the guidelines.

Acknowledgements

The study was commissioned by the Environment Division of the Inter-American Development Bank with Luis Garcia, project leader and Gil Nolet, legal specialist. The study was financed through the core funds of the Netherlands Development Agency (NEDA) under contract number IDB (ATN/NE - 5661-RG)

Development activities have been carried out by a team of IHE experts and local consultants: Mr. Paul van Hofwegen (Project Coordinator), Mr. Frank Jasper, Prof. Dr. ir. Hubert Savenije. Local consultants either are IHE alumni in Water Resources Management or liaised with IHE through its network: Mr. Orlandino Arteaga Toledo (Guatemala, private consultant), Mrs. Michelle Watts (Jamaica, Water Resources Authority), Mr. Edgar Quiroga Rubiano, Mrs. Mariela Garcia, Mr. Anibal Valencia and Mrs. Claudia Nieto (Colombia, Universidad del Valle Cali - Instituto CINARA).

The development and testing of the framework and the guidelines would not have been possible without the cooperation of all the persons interviewed and participants of the workshops in Guatemala City, Kingston Jamaica and Bogota Colombia. The assistance from the local IDB offices in organizing the workshops and providing venue is highly appreciated.

Delft, The Netherlands
December 1998

Abbreviations and Acronyms

ADB   Asian Development Bank
BOOT   Build-Own-Operate-Transfer
BOT   Build-Operate-Transfer
CINARA   Instituto de Investigacion y Desarollo en Agua Potable, Saniamiento Basico y Conservacion del Recurso Hidrico, Universidad del Valle, Cali, Colombia.
GO   Government Organization
IDB   Inter-American Development Bank
IHE   International Institure for Infrastructural, Hydraulic and Environmental Engineering Delft, The Netherlands.
IWRM   Integrated Water Resources Management
LAC   Latin America and the Carribean Countries
MIS   Management Information System
NEDA   Netherlands Development Agency
NGO   Non-Governmental Organization
O&M   Operation and Maintenance
PLC   Public Limited Company
PSP   Private Sector Participation

Summary

The Inter-American Development Bank (IDB) developed a strategy for Integrated Water Resources Management (IWRM) which aims to help borrowing member countries to shift from a sectoral, development-based focus to an integrated, management-based approach. One key principle of this Strategy is an increased emphasis on institutional issues and capacity building.

Integrated water resources management (IWRM) means decision making on development and management of water resources for various uses taking into account the needs and desires of different users and stakeholders. IWRM focuses at interests in use, control or preservation of water systems and their sustainability.

To pursue IWRM, two situations need to be assessed (Lord and Israel 1996, IDB 1997). The first is the context in which policy is pursued and programs developed. It is composed of actors, whose actions are shaped by the environment, whether natural or manufactured, and by rules. The rules define the relationship between actors and the environment. They describe for example, how cost and benefits are allocated among actors, how authority is distributed, who decides and how, or who has access to what information. The second is the level at which actions and decision making occurs and, by extension, where integration occurs. For this purpose, three functions are considered: the operational or water use function, the organizational or water resource management function, and the constitutional or water policy and law function.

The operational function is focussed at use or control of water for specific purposes to fulfill specific needs and demands. These include water supply and sanitation, irrigation and drainage, flood protection, hydropower, industrial supplies, tourism and recreation, fisheries, navigation and the preservation or rehabilitation of ecosystems.

To reduce the problems and conflicts between these different uses and users a platform is required for coordination of water use and water allocation. Solving these problems also requires establishment or changes of water use rules. This is the organizational function. It involves coordination, planning, decision making and policing of water use and users in water systems (river basins, aquifers).

To make the organizational function possible an enabling environment has to be created. This requires water policies, institutional development policies, including human resource development and normative and executive legislation. This is the constitutional function. These higher level actions are important because ineffective rules, accountability and policing mechanisms assure that water use and control problems cannot be solved (IDB 1997).

IWRM is a process of assignment of functions to water systems, the setting of norms, making allocations for use, enforcement (policing) and management. It includes gathering information, analysis of physical and socioeconomic processes, weighing of interests and decision making

related to availability, development and use of water resources. IWRM requires therefore
*   A platform for weighing of all relevant interests and decision making on use of water and water systems in the river basin,
*   This platform should represent all interests and be under governance of government to protect the interest of society;
*   This platform should have decision, control and sanctioning powers;

A minimum set of conditions should be met to allow such IWRM platforms to operate successfully. These conditions are related to constitutional, organizational and operational functions. For all these functions it is required that the respective authorities have the mandate and the resources (financial and human) to carry out their tasks in development and implementing IWRM.

IWRM requires from the constitutional function a system that
*   enables effective development and implementation of laws and regulations,
*   enables effective constitution and development of relevant institutions,
*   regulates decision making based on interests of all stakeholders,
*   enables all stakeholders to participate in decision making,
*   provides quantitative and qualitative standards for use,
*   provides quantitative and qualitative standards for effluents,
*   enables and regulates effective control and sanctioning of violations,
*   enables implementing agencies to take the necessary steps in order to secure and conserve the resource,
*   provides effective and transparent accountability mechanisms.
*   provides sufficient capable people to meet the IWRM demands of policy making, adapting legislation and all other activities
*   enables and regulates private sector participation

The basic function at organizational level is to coordinate between the different stakeholders and to decide on the different uses of water. For an organizational function to be carried out effectively it requires
*   a decision making capacity on (sub-) river basin level that reflects the interests of different uses and users,
*   a clear regulatory framework with norms and standards for decision making,
*   a system that provides reliable information on the availability, use and quality of surface and ground water in the (sub-) basin,
*   a system that allows analysis of several scenarios for development and use of water at basin level,
*   an effective and transparent accountability mechanism,
*   power to control and sanction violations,
*   sufficient capable people to meet the IWRM demands on planning and management, and control.

Effective operational functioning within an IWRM context requires a management system that responds to societal needs. This means that for water services the system should enable, provide or regulate:
*   effective control of the service providers by users/clients and the IWRM Platform
*   representation of clients' interests at and by the managing agency
*   cost recovery by the service provider.
*   negotiations between the managing agency/service provider and its clients on the level of

service it provides and recovery of its associated cost,
- assessment of the demands, actual use and availability of water (quantitative and qualitative)
- power at the service provider to control and sanction violations
- sufficient capable people to meet the IWRM demands for planning, development and management of the services provided.
- a system that allows market incentives to make most economic use of water through participation of the private sector.

To arrive at this set of requirements at all these functional levels a process of capacity building for IWRM is needed in a service oriented environment. An analytical framework is proposed to make a proper assessment of required capacity building interventions. This framework is based on the development process to come from an identified 'present water resources management situation' to some 'desired integrated water resources management situation'. The desired IWRM situation is a compromise between the present and the ideal IWRM situation as complete introduction of IWRM is unrealistic and maybe undesirable to expect. This compromise will be the result of a negotiation process in which policy makers, water resources and water utility managers and stakeholders are involved. The outcome will be determined by technical, financial and political attainability under prevailing socioeconomic conditions. With changing conditions the desired IWRM situation will change.

This process consists of the following major steps:
- assessment of the present situation and trends,
- formulation of a desired IWRM situation based on an ideal or eventual IWRM situation;
- formulation of interventions to arrive at the desired IWRM situation,
- establishment of a monitoring system to see whether the interventions are being carried out properly and whether they really contribute to the achievement of the IWRM goals.

A proper assessment of the institutional situation requires a good understanding of the physical conditions, the important stakeholders and their relationships, the current problems and envisaged solutions. A draft set of guidelines for assessment of the institutional framework was developed and tested for the three functional levels (constitutional, organizational and operational) in three pilot cases: Guatemala, Jamaica and Colombia. Based on the experience gained by the local consultants and the outcome of local workshops the following steps for assessment by an IDB expert team are recommended:

1. Experts provide a (desk study) report about the existing water management situation combined with registered problems (quantity, quality and environment). The report serves as a general background document for all the following steps and has to be distributed accordingly.

2. Under auspices of the IDB a representative team is formed to identify and select relevant stakeholders from the categories: water policy makers, water managers, water service providers, water using agencies, water using groups, water users and other potential interest holders at constitutional, organizational and operational functional levels.

3. Experts carry out an elaborate procedure of interviewing the selected stakeholders applying the guidelines for interviews. These guidelines are as a questionnaire that contain questions about the existing situation, specific problems and their causes and the desired situation according to the interviewed stakeholder. During this interview previously

overlooked stakeholders can be identified. The interviews are analysed according to the matrix model.

4. The results of the interviews are described in a report and will be sent to the interviewed stakeholders together with the background document. These stakeholders should also be invited to the following workshops.

5. The first workshop to which all the relevant stakeholders are invited deals with the assessment of the existing water resources management situation and problem identification as perceived by the respective stakeholders.

6. The second workshop (one to three months after the first workshop) will elaborate extensively on the principles of integrated water resources management. The outcome will be the formulation of a desired water resources management situation in that specific country and the set of interventions that will be needed to achieve that.

7. Experts draft a preliminary country document comprising:
    -   assessment of the existing water management situation
    -   a complete problem inventory
    -   the desired water resources management situation
    -   a proposed set of general interventions needed to reach the desired situation and if feasible, suggested specific interventions to be included in Bank operations

8. The draft country report is distributed and a thorough procedure for collecting comments is followed.

9. Experts draft a final country report offered to the IDB for endorsement and inclusion into the strategy and/or into specific water related projects for the specific country.

10. The IDB designs a monitoring procedure to follow whether the interventions are taking place and whether the envisaged results are achieved.

To help the process of assessment, a set of tools has been developed as guidelines for interviews of stakeholders, evaluation matrices and water use flow diagrams.

The guidelines can be applied in different stages of the project cycle. Its use should result in an agreement on what the actual problems and conflicts are and what interventions are needed to overcome these. These interventions can be included in the project design. These projects can be sectoral on regional or (sub-) basin scale or sub-sectoral and on local scale.

The application of the guidelines is most effective in sector wide institutional change and development programs. The guidelines can also be applied in relation to local physical infrastructure projects. These projects should be of a scale that different stakeholders will be influenced and conflicts of interests are foreseen on local and (sub-)basin scale.

The guidelines can be applied at different levels of scale: sub-basin, basin and national level. For whatever level of scale these guidelines are applied, identifying and engaging all relevant stakeholders at the three functional levels is crucial. This exercise cannot be replaced by a desk study as the experience of the Chile study showed.

Engaging independent local experts is advisable. They should preferably not come from within the government

The use of the guidelines involves use of time, financial and people. The amount of these resources required depends on the scale and level of the project to which the process is attached. It also depends on the awareness on IWRM among the stakeholders, the political desire to introduce IWRM, the availability of information and the size and accessibility of the (project) area envisaged. The main cost items are the local and international specialist fees, travel and living expenses of the specialists, the organisation of workshops and the travel and living cost of the participants of the workshops. The Bank normally does not finance travel and per diem expenses for workshop participants.

Often the scale of the process will not be sufficiently known. In such case the first step of the process (inventory) can be separated from the remainder. The outcome of step one should then include a cost estimate for one cycle of the process.

CHAPTER 1

The Analytical Framework

1.1 INTRODUCTION

The Inter-American Development Bank (IDB) developed a strategy for Integrated Water Resources Management (IWRM) which aims to help borrowing member countries to shift from a sectoral, development-based focus to an integrated, management-based approach. One key principle of this Strategy is an increased emphasis on institutional issues and capacity building. This requires an analytical framework for the assessment of the institutional setting for integrated water resources management that the bank could use for the incorporation of capacity building considerations in future IDB water related projects.

The framework and the guidelines are made for project teams, bank officers and government agencies to facilitate the process of project formulation and monitoring. The framework should improve the integration of different steps in the project cycle of project managers and bank field offices and generate more involvement.

1.2 FUNCTIONS AND FUNCTIONAL LEVELS IN IWRM

IWRM means decision making concerning development and management of water resources for various uses. In this decision making process it takes into account the needs and desires of all the different uses, users and stakeholders. To pursue IWRM, two situations need to be assessed (Lord and Israel 1996, IDB 1997). The first is the context in which policy is pursued and programs developed. It is composed of actors, whose actions are shaped by the environment, whether natural or manmade, and by rules. These rules define the relationship between actors and the environment. They describe, for example, how cost and benefits are allocated among actors, how authority is distributed, who decides and how, or who has access to what information. The second is the level at which actions and decision making occurs and, by extension, where integration occurs. For this purpose, three functions are considered (figure 1): the operational or water use function, the organizational or water resource management function, and the constitutional or water policy and law function.

The operational function is focussed at use or control of water for specific purposes to fulfill specific needs and demands. These include water supply and sanitation, irrigation and drainage, flood protection, hydropower, industrial supplies, tourism and recreation, fisheries, navigation and the preservation or rehabilitation of ecosystems.

To minimize the problems and conflicts of these different uses and users, coordination of water use and water allocation is required. Solving these problems also requires establishment or changes of water use rules. This is the organizational function. It involves coordination, planning, decision making and policing of water use and users in water systems (river basins, aquifers)[1].

To make the organizational function possible an enabling environment has to be created. This requires water policies, institutional development policies, including human resources development and normative and executive legislation. This is the constitutional function. These higher level actions are important because ineffective rules, accountability and policing mechanisms assure that water use and control problems cannot be solved (IDB 1997).

IWRM requires a good performance at al these functional levels. Development towards IWRM therefore needs to address these levels in a holistic way. These development efforts are called capacity building. To make a proper assessment of required capacity building interventions the following framework is proposed.

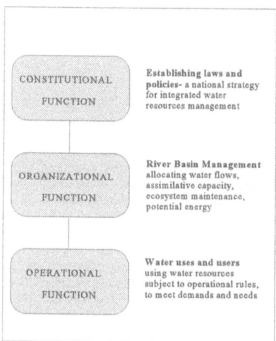

Figure 1. Functions in integrated water resources management.

1.3   THE ANALYTICAL FRAMEWORK

The analytical framework is based on a cyclic development process to come from an identified present water resources management situation to some desired integrated water resources

---

[1]   This concept was described by Lord and Israel (1996). In their terminology they used constitutional level, organizational level and operational level. The use of the term level lead to much confusion as it implicitly refers to administrative levels like national level for constitutional functions, river basin level for organizational functions and project or scheme level for operational functions. Implicit this assumes a separation of these functions among the levels. In practice one organization can have different functions which can be implemented at one and the same level. The organizational function however, can only be carried out by an independent organization that is not in charge of operational (project) functions.

management situation. The desired IWRM situation is a compromise between the present and the ideal IWRM situation as an instantaneous complete introduction of IWRM is unrealistic and maybe undesirable to expect. This compromise will be the result of a negotiation process in which policy makers, water resources and water utility managers and stakeholders are involved. The outcome will be determined by technical, financial and political attainability under prevailing socioeconomic conditions. With changing conditions the desired IWRM situation will change. This process contains the following major steps (figure 2):

- Assessment of the present situation and trends,
- Formulation of a desired IWRM situation based on an "ideal" or eventual IWRM situation,
- Formulation of interventions to arrive at the desired IWRM situation,
- Establishment of a monitoring system to see whether the interventions are being carried out properly and whether they really contribute to the achievement of the IWRM goals.

The analytical framework and the guidelines are developed for the three functional levels: constitutional, organizational and operational.

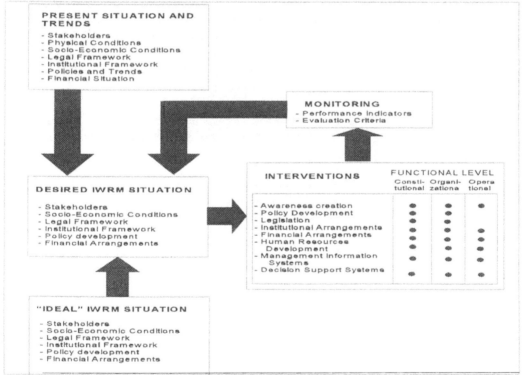

Figure 2.   Analytical Framework for the assessment of the institutional setting and capacity building requirements for integrated water resources management

1.4   METHODOLOGICAL STEPS

A proper assessment of the institutional situation requires a good understanding of the physical conditions, the important stakeholders and their relationships, the current problems and envisaged solutions. Local consultants used a draft set of guidelines for assessment of the institutional framework in three pilot cases: Guatemala, Jamaica and Colombia. A desk study was made of the situation in Chile. Based on the experience gained by the local consultants and the outcome of local workshops during the test cases the following steps for assessment by an IDB expert team are recommended:

1.   Experts provide a (desk study) report about the existing water management situation combined with registered problems (quantity, quality and environment). The report serves as a general background document for all the following steps and has to be distributed accordingly.

2.   Under auspices of the IDB a representative team is formed to identify and select relevant stakeholders from the categories: water policy makers, water managers, water service providers, water using agencies, water using groups, water users and other potential interest holders at constitutional, organizational and operational functional levels.

3.   Experts carry out an elaborate procedure of interviewing the selected stakeholders applying the guidelines for interviews (appendix I). These guidelines are presented as a questionnaire that contain questions about the existing situation, specific problems and their causes and the desired situation according to the interviewed stakeholder. During this interview previously overlooked stakeholders can be identified. The interviews are analysed according to the matrix model. (Appendix II).

4.   The results of the interviews are described in a report and distributed with the background document among the interviewed stakeholders. These stakeholders should also be invited to the following workshops.

5.   The first workshop to which all the relevant stakeholders are invited   deals with the assessment of the existing water resources management situation and problem identification as perceived by the respective the stakeholders.

6.   The second workshop will take place one to three months after the first workshop. It will elaborate extensively on the principles of integrated water resources management. The outcome of the workshop will be the formulation of a desired water resources management situation in that specific country/basin/sub-basin and the set of interventions that will be needed to achieve that.

7.   Experts draft a preliminary country document comprising:
        - assessment of the existing water management situation
        - a complete problem inventory
        - the desired water resources management situation
        - a proposed set of general interventions needed to reach the desired situation and, if feasible, suggested specific interventions to be included in Bank operations.

8.   The draft country report is distributed and a thorough procedure for collecting comments is followed.

9.   Experts draft a final country report that will be offered to the IDB for endorsement and inclusion into the strategy and/or into specific water related projects for the specific country.

10.   The IDB designs a monitoring procedure to follow whether the interventions are taking place and whether the envisaged results are achieved.

Before the description of the institutional assessment and capacity building processes describing the general IWRM principles and their consequent requirements at the different function levels is necessary. The use of the framework and the methodological steps is then further elaborated in Chapter 5 on process and tools for institutional framework assessment.

CHAPTER 2

Integrated Water Resources Management - General Principles

2.1 AIM AND DEFINITION

International awareness about the importance of water resources management is growing. Originally the approach was very sub-sectoral, mostly in relation to water supply, sanitation and irrigation. There is however, a growing consensus that Integrated Water Resources Management (IWRM) is necessary for sustainable resource use for all the sub-sectors and to protect the environment.

The aim of IWRM is to discard with the one-sided management perspective of single interests of one sub-sector by one government agency and to strive for a participatory multi-sided management perspective of all interests in management of water resources. IWRM therefore takes account of all natural aspects of the water resources, all sectoral interests and stakeholders, the spatial and temporal variation of resources and demands, relevant policy frameworks and all institutional levels.

IWRM can be defined as follows: (after Verhallen ea., 1997):
> *Integrated water resources management is the management of surface and subsurface water in qualitative, quantitative and ecological sense from a multi disciplinary perspective and focussed on the needs and requirements of society at large regarding water.*

In practice this implies:
a.   recognition of the fact that ecologically healthy functioning water systems are the basis for sustainable use by man, flora and fauna;
b.   recognition of the fact that by management of these systems all interests need to be considered in the functioning of the water systems and that regulation is required to guarantee its sustainable use.
c.   recognition of the fact that interests can best be represented in a platform with planning, coordination, decision and policing powers in which all relevant stakeholders are represented.

2.2 WATER SYSTEMS

The object of IWRM is the water system that can be defined as a geographically defined coherent and functioning entity of surface water, ground water, green water[2], fossil water, river and stream beds, banks and technical infrastructure including all accompanying physical, chemical and biological characteristics and processes. The boundaries of such a water system are determined based on a functional coherence between hydrological, morphological and ecological aspects. For development and management of water systems their use functions play an important role. If these use functions are considered during the processes of planning, development and management of such a water system one can speak of a water systems approach. This approach aims at matching the desires of society regarding the use of water with the functioning of water systems through technical and legislative interventions. (Omgaan met water, 1985)

The system boundaries are to be determined such that the system under consideration can be considered isolated from its environment. The environment can influence the system but the influence of the system on the environment is small. The scale of the system is determined by the purpose or the aim of the system. In IWRM context we speak system boundaries coinciding with river basins and sub-basins.

Society with its needs, demands and desires and the role of government to match these with the potential of the water systems are part of the environment of water systems. Based on these potentials the interests of society have to be weighed and decisions have to be made on specific use functions of these systems or part of this. Systems can then be managed using water quantity and quality norms or standards for specific functions in specific (sub-)systems directed towards preservation, improvement or development of the potentials of that (sub-) system in a sustainable way.

2.3 NEED FOR INTEGRATED WATER RESOURCES MANAGEMENT

Actions to use or control water for specific purposes are aimed at security, social well being, economic gain and the preservation of ecosystems. These use and control activities can create problems that may be classified as externality, open access, public interest and scarcity problems (Lord and Israel, 1996).

Externality problems exist when actions of one party affect the well being of a second party, and the first party cannot itself gain by considering this effect and modifying its behaviour accordingly. Open access problems exist when the use of the resource is open to all, and when the rate of use of that resource affects the amount that can be used. Public interest problems relate to the necessity to provide a particular good to all in equal amounts. No one can be excluded from consuming it, and the cost of providing it to one is as great as the cost of providing it to all. The problem is that these goods are likely to be under provided because no one will undertake to

---

[2]   Falkenmark (1995) developed the concept of blue and green water. Blue water is that part of rainfall that ultimately ends up in the water conveyance system: rivers, lakes and aquifers. Green water is that part of rainfall that enters into the unsaturated soil where it is consumed directly for evaporation and transpiration.

produce them, since they cannot be withheld from others, thus cannot be sold to make profit. These goods thus must be provided by government. Scarcity problems exist when the users want more of a good than the quantity available at a given price. Economic markets handle scarcity by allowing competition, in which those with the most purchasing power, and to whom the resource is most valuable, will bid it away from others. To safeguard the low-income strata of society and the environmental needs, the negative effect of scarcity is commonly dealt with by non-market institutions such as river basin councils or the government.

Solving these problems requires establishment or changes of water use rules that must occur at water resource level. Creating an effective set of water resource management rules requires action at the water policy and law level. These higher level actions are important because ineffective rules and ineffective accountability and policing mechanisms assures that water use and control problems cannot be solved.

2.4   SUSTAINABLE USE OF WATER RESOURCES

Presently the definition most often used of sustainable development is: the ability of the present generation to use its natural resources without putting at risk the ability of future generations to do similarly. Sustainable development is making efficient use of our natural resources for economic and social development while maintaining the resource base and environmental carrying capacity for coming generations. This resource base should be widely interpreted to contain besides natural resources: knowledge, infrastructure, technology, durables and people. In the process of development natural resources may be converted into other durable products and so remain part of the overall resource base.

Water resources development that is not sustainable is ill-planned. In many parts of the world, fresh water resources are scarce and usually finite. Consequently, there are many ways to jeopardize the future use of water, either by over exploitation (mining) of resources or by destroying resources for future use (e.g. pollution). Besides physical aspects of sustainability there are social, financial and institutional aspects. The following aspects of sustainability are distinguished (Savenije 1997):
- technical sustainability (balanced demand and supply, no mining)
- financial sustainability (cost recovery)
- social sustainability (stability of population, stability of demand, willingness to "pay")
- economic sustainability (sustaining economic development or welfare and production)
- institutional sustainability (capacity to plan, manage and operate the system)
- environmental sustainability (no long-term negative or irreversible effects).

2.5   INTERESTS

There are many interests in water. With interests in water is meant the benefit obtained or preserved by individuals, groups or nations with the presence, use and control of water. Interests can be classified of those of the first and of the second order. Interests of the first order are essential conditions for life in that water system: human, animal and plant. Interests of the second order are those that can be prioritised after being weighed on their economic, ecological and social values.

Government has the "care function" as for management of water resources. First order interests are interests of society and therefore require to be represented by government. Second order interests are interests of individuals, groups or a part of society and can best be represented by their stakeholders.

First and second order interests are different in place because of different physical, hydrological, cultural and socioeconomic conditions. As development goes on, especially second order interests will change. This means that interests are site specific and time specific and so site specific approaches are warranted.

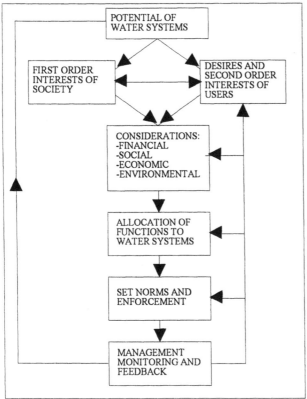

Figure 3   The process of assignment of functions to water systems and the setting of norms, enforcement and management (after Verhallen et al, 1997.)

Some common interests are:

- drinking water for population
- agriculture, fisheries, livestock
- safety against floods and inundations
- industry (process, cooling)
- recreation
- nature and landscape
- conservation of ecosystems
- transport and navigation
- flushing

In water resources allocation there is general agreement that the supply of water for basic human needs has priority. In this respect the equity principle plays a major role. Another priority is the requirement to maintain essential life support ecosystems. These can be considered first order priorities. All other needs for industry, agriculture or other societal needs should be prioritised according to socioeconomic criteria, by which water is considered an economic good. Although cost-recovery and economic pricing are overriding principles, pricing and tariff regulations within sub-sectors are considered necessary where equity or social well-being are at risk and environment is endangered.

Another clear consensus is the need for adequate participatory approaches to planning and management, and mechanisms for accountability and democratic control. This is closely related to the principle of decision-making at the lowest appropriate level (subsidiarity), which also implies that some decisions (for instance on the sharing of international waters) should be taken at the highest level. In that case, mechanisms of democratic control and stakeholder participation clearly operate at the highest level of government.

The river basin is the logical unit for water resources management. In many cases this has led to the decentralisation of management to river basin level. But one should not forget the role of central government. River basin management is largely an operational matter, whereby water allocation, water quality management, cost recovery and stakeholder involvement are essential components. However, the river basin authority is not a legislator and not responsible for policy making and the setting of objectives and constraints to operational management.

Central government has an important role in IWRM in policy making, legislation, strategic planning, establishment of the appropriate legal and institutional framework, capacity building, and supervision of decentralised and privatised institutions in water resources management. In addition, government should provide the protocols for information exchange (on water resources, water use and infrastructure), should provide adequate databases required for strategic planning and should prepare integrated river basins plans in response to its policy guidelines and constraints.

2.6   THE "IDEAL" IWRM SITUATION

In the analytical framework an "ideal" IWRM situation is formulated that should give direction to the process of integration of the management of water resources. In an ideal IWRM situation the water resource is managed on (sub-)basin level in a sustainable way. The water related interests of all stakeholders are considered in decision making on water use. All stakeholders are aware of the potential of the water source and the impact of their use on the other stakeholders. Decisions on water use and associated cost of service provision are made in a participatory manner according criteria agreed and accepted by all stakeholders. Implementation of IWRM is done at the least cost in a transparent way with effective accountability mechanisms in place.

Before IWRM can be successfully carried out, a set of institutional conditions must be met. These requirements are at the three functional levels and will be discussed in the following chapter.

CHAPTER 3

IWRM Requirements for the Functional Levels

IWRM is a process of assignment of functions to water systems, the setting of norms, enforcement (policing) and management. It includes gathering information, analysis of physical and socioeconomic processes, weighing of interests and decision making related to availability, development and use of water resources. This means that IWRM requires:

- a platform for weighing of all relevant interests and decision making on use of water and water systems in the river basin;
- this platform should represent all interests and be under governance of government to protect the interest of society at large;
- this platform should have decision, control and sanctioning powers;

A minimum set of conditions should be met to allow such IWRM platforms to operate successfully. These conditions are related to constitutional, organizational and operational functions. For all these functions it is required that the respective authorities have the mandate and the resources (financial and human) to carry out their tasks in development and implementing IWRM.

3.1   THE CONSTITUTIONAL FUNCTION: WATER POLICY AND LAW

The main purpose of the constitutional function is to create an enabling environment for the IWRM platform with appropriate policy and legal frameworks which gives the boundary conditions for effective implementation of the organizational and operational functions. Constitutional functions include policy development based on clear principles, development of normative and executive legislation and development of human resources development strategies. An important aspect to be arranged at this functional level is the degree level of participation of the private sector in all three the functional levels.

3.1.1   Principles for Water Policy

In recent years, many governments, national and international institutions and agencies have reached a consensus on principles to govern the formulation of a national water policy. For the LAC countries these principles (see box 2) evolved from The Dublin conference (1992), the

Global Water Partnership (1996), the San Jose Declaration (1996) and the Declaration of Buenos Aires (1996).

As an example of principles for development of water policies the case of the Asian Development Bank is given. These principles are based on the outcome of a regional workshop and are as such region specific. However, the principles show a high level of agreement with the IDB Strategy (1997) and the Worldbank strategy (1993). The Asian Development Bank (ADB, 1996) identified seven principles for effective water policy. The first three refer to essential water sector functions and the second four are cross cutting principles for successful water sector activities:

Principles for essential water sector functions:

1. National water resources development and management should be undertaken in a holistic, determined and sustained manner to meet national development goals and protect the environment

2. Planning, development and management of specific water resources should be decentralized to an appropriate level responding to basin boundaries.

3. Delivery of specific water services should be delegated to autonomous and accountable public, private, or cooperative agencies providing measured water services in a defined geographical area to their customers and/or members for an appropriate fee.

Box 1: Dublin principles and associated concepts for water resources management

Dublin Principles
*   Water is a finite and vulnerable resource, essential to sustain life, development and the environment, which should be managed in an integrated manner;
*   Water resources development and management should be based on a participatory approach, involving all relevant stakeholders, and making decisions at the lowest appropriate level (subsidiarity);
*   Women play a central role in the provision, management and safeguarding of water
*   Water has an economic value in all its competing uses and should be recognised as an economic good, taking into account affordability and equity criteria.

Associated key concepts
*   Integrated water resources management, implying:
    - an intersectoral approach
    - representation of all stakeholders
    - all physical aspects of the water resources
    - sustainability and environmental considerations
    - trans boundary water resources issues
*   Sustainable development, sound socio-economic development that safeguards the resource base for future generations
*   Emphasis on demand driven and demand oriented approaches
*   Aim to help people, especially the poor and other vulnerable groups, to benefit from improved water resources management while safeguarding the environment;

Crosscutting principles for successful water sector activities

4. Water use in society should be sustainable-with incentives, regulatory controls, and public education promoting economic efficiency, conservation of water resources, and protection of the environment - within a transparent policy framework.

5. Shared water resources within and between nations should be allocated efficiently for the mutual benefits of all riparian users.

6. Water sector development activities should be participatory and consultative at each level, leading to commitment by stakeholders and action that is socially acceptable.

7. Successful water sector development requires a commitment to sustained capacity building, monitoring, evaluation, research and learning at all levels, to respond effectively to changing needs at the national, basin, project, service entity and community level.

Water sector development and management should therefore be undertaken in a holistic, determined and sustained manner to meet national development goals while protecting the environment within a transparent policy framework without which investments cannot fulfill their potential (Ait Kadi, 1997). A policy statement is required as an expression of a long term vision and commitment at the highest level of government. Putting the idea of holistic management into practice also means that governments need to strengthen existing or create new institutions capable to carry out integrated water management. They need to assign high priority to capacity building, including human resources development. There is also a need to put in place a coherent set of incentives and regulatory controls to support national water policies. Investments are required to accomplish this. Moreover, investments are required to mobilise additional water resources, improve water use efficiency and to restructure the economy of the country and its consumption patterns away from wasteful and low value water uses.

3.1.2 Legal Frameworks

A legal framework serves two purposes:
*   to ensure that the agreed policies will be implemented (normative legislation),
*   to provide the proper tools for implementation (executive legislation).

Normative legislation thus translates the policy principles into laws. In development of legislation a distinction has to be made between first and second order policy principles.

Policy Principles of the First Order

First order policy principles are principles that are not negotiable and that cannot be overruled by other principles such as:
*   Water is a basic need. No human being can live without a basic volume of water. Most life processes depend on availability of at least a minimum amount of water.
*   Water resources systems can only persist in a natural environment that can regenerate (fresh) water of sufficient quality. Therefore, only sustainable water use can be allowed to the extent that future generations still can use the same resource base and can generate water resources in a similar way as present generations.

These principles are inalienable and entering them in to the Constitution is advisable.

Policy Principles of the Second Order

Policy principles of the second order may change from time to time and from place to place depending on the consensus of the specific society. However, the following principles are accepted as a result from extensive international development processes:

- Water has to be distributed in an equitable way, so that an optimal benefit results for an as large as possible group of stakeholders.
- Water is an economic good and should be priced under all circumstances to reach sustainable economic development. Principles like 'users and polluters pay' have proven their value.
- Water resources management should be based on a participatory approach, involving all relevant stakeholders.
- Water resources management tends to be more efficient when carried out on hydrological boundaries applying integrated planning techniques, based on demand driven and demand oriented approaches.
- Water resources management tends to be more effective when decisions are made at the lowest appropriate level.
- Water is an inseparable entity. There is no basic difference between managing underground, surface, atmospheric or any other water.

These principles are negotiable and are advised to be entered in National Laws.

Legal Tools

When setting up a suitable legal framework for implementation of IWRM the following aspects may have to be handled in executive legislation:

- International catchment management
- Integrated planning
- Water rights or water permit system
- Transfer and mobility of water rights and permits
- Arbitration and appeal
- Control, policing and sanctioning
- Institutional development
- Financial accountability
- Delegation and decentralization
- Participation of water users and stakeholders
- Commercialization and privatization
- Standards for water quality, emissions and safety

When society is getting more complex, with an increasing and more diverse water use, a legal framework for water resources management needs more differentiation and flexibility. Normally this results in the functional differentiation to a constitutional, organizational and operational level. Each of these levels have their specific legal tools which of course should be provided by the constitutional level. The following guideline for legal tools assessment could be instrumental:

Constitutional Function: Convention of international organization
Bilateral or multilateral treaty or agreement
Constitution
National Legislation
National Plans

Organizational Function: State Regulations or Plans
Ministerial Regulations
Regulation or plans of functional public body like national water authority, (sub-) catchment authority
Provincial regulation or plan

Operational Function: Sub-catchment-, district, town regulations
Bye-laws of semi-public or private water users organizations etc.

The boundaries between the functional levels are not sharp and the function of the legal tools may vary.

The complexity of water management induces a need for delegation of regulating authority to the decentralized organizational and operational levels. For instance, it has become unpractical to pose water quality or emission standards by law. This is done through a more flexible or quickly adaptable legal tool. Usually it can be stated that the more detail is required, the more decentralized and flexible the legal tool has to be. Further, the more complex the situation, the more emphasis should be on water users or stakeholder participation, demand management etc. meaning management at the lowest appropriate level.

In an increasingly complex water management situation, also the need for more coherent planning mechanisms arises. A cornerstone of IWRM is the development of a rational system of integrated planning, preferably at all functional levels.

Three types of plans can be distinguished: policy (strategic) plans, development plans and operational/management plans. It is essential that these types of plans have as much as possible an integrated character. In practice this means that the plans are dealing with all relevant aspects (water quantity, water quality, environmental setting) and all relevant water systems (surface, subsurface, (un)saturated zones, etc.). It is becoming a general practice to develop management and development plans on hydrological boundaries for specific (sub)catchments. It is essential that any planning is done by or in close consultation with a full representation of water users and other (potential) stakeholders through the relevant platform(s).

Generally, policy plans are developed at the constitutional functional level, development plans at the organizational functional level and operational/management plans at the operational functional level. This, however, may vary from country to country. Mixtures of plans are common. Especially, development plans are often clustered or combined with operational/management plans. Important is that institutional mechanisms are present to finetune the consistency and interdependency of the plans.

3.1.3   Private Sector Participation

Private sector participation (PSP) in the water sector is most developed in the hydropower and water supply and sanitation sub-sectors. In other sub-sectors PSP is often limited to involvement of contractors and consultants for temporary projects. To understand the degree of involvement or participation of the private sector six basic modes of water (supply) sector organizations are presented in table 1. The scheme offers a useful shorthand for the discussion of complex water utility organisational issues following below.

Table 1   Water sector organisation: six basic modes (Braadbaart and Blokland 1998)

Mode of organisation	Who owns the infrastructure?	Who operates the infrastructure?	Legal status of operator	Who owns the shares?
Direct public/ local	Local (municipal) government	Local (municipal) administration	Local (municipal) department	Not applicable
Direct public/ supra-local	National or state/ provincial government	National or state government administration	National or state government department	Not applicable
Corporatised Utility (corporation/authority/ board)	Government or Utility	The corporatised utility	Para-statal, usually defined by special law	Not applicable
Public-owned Public Limited Company (PLC)	Government or PLC	A PLC as permanent concessionaire	Public Limited Company	Local/provincial government
Delegated private	Any combination of government agencies	Government and temporary private concessionaire	Public Limited Company	Private shareholders
Direct private	Private agents	Private company	Public Limited Company	Private shareholders

Source: EUREAU (1992) and Sector and Utility Management Group data bank IHE (Franceys, 1997)

The adapted scheme distinguishes six basic modes of water utility organisation in terms of:
*   ownership of the utility infrastructure, that is headworks, treatment plant, network, and other assets;
*   the identity of the system operator;
*   the legal status of the system operator; and
*   the ownership of the shares of the operating company, where applicable.

The terminology introduced here is interchangeable with terms used across the water supply industry. Thus, the direct public/local management mode comprises for example the municipal waterworks departments found in countries as diverse as Indonesia, the United States, and Spain. It has also recently grown into a dominant form in the transition countries of Central and Eastern Europe. This former central planning has shifted from direct supra-local to direct local government, with some contemplating a further downgrading of central government involvement by introducing delegated private management.

The direct/supra-local management option refers to large government departments, at state/province or national level charged with the management of multiple schemes that serve most municipalities for water supply or irrigation schemes on state or national scale. This form is, e.g. found in the State Public Health Engineering Departments and irrigation departments in India.

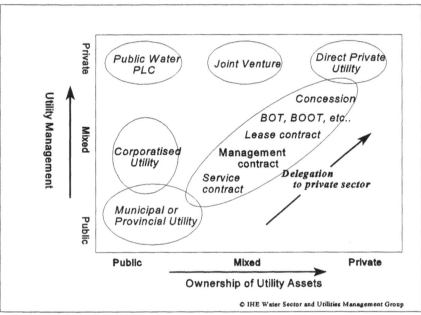

Figure 4 Degrees of Private Sector Participation in terms of Utility Management and Ownership of Utility Assets (Braadbaart and Blokland, 1998)

Under corporatised utility the prevalent management situation is described in most developing countries. Here one finds large organisations, e.g. the State Water Boards in India, the Ghana Water Supply and Sewerage Corporation, or the Provincial Waterworks Authority of Thailand, responsible for water supply and assorted other services on a country or state scale.

Delegated private management describes what is known as the French system of out contracting construction and O & M activities to private firms. Delegated private management is also the management option currently favoured by the World Bank. In the developing world, it can be found mainly in mega-cities, e.g. Buenos Aires and Manila.

The direct private mode describes what is also known as the British model. More precisely it refers to the current situation in England and Wales, whose water utilities are both privately owned --shares are traded on the Stock Exchange-- and privately managed.

The public-owned Public Limited Company (PLC) refers to a mode of organisation where both the utility's infrastructure and the shares of the water company are owned by local and provincial government representatives. The operator (PLC) is an autonomous for-profit organisation falling under commercial law. The public water PLC is crucially different from French delegated private management in that the operator is owned by public rather than private shareholders. Furthermore, the public water PLC is mostly a permanent concessionaire where its French counterpart is a temporary concession-holder. The public water PLC also differs from direct public management, and it does so in two important respects: consumer influence and autonomy. First, under the public water PLC structure the utility's consumers have a direct say in strategic decisions, e.g., their representatives must approve of the annual budget, an investment plan, or a proposal to change the tariff. Consumer interests may be exerted in various ways. In the Philippines, five representatives of local interest groups (business, women, and so on) form a Governing Board that meets with the General Manager of the water utility on a regular basis. Umgeni Water board in Kwazulu-Natal province, South Africa, works under a similar, if larger, governing board. Under the Dutch system, consumer delegates exercise their power through the Board of Directors and the annual Shareholders Meeting.

Second, unlike direct public management, the public water PLC is always an autonomous for-profit entity. Unlike the municipal waterworks of direct public/local management, it does not form part of the administrative apparatus of a town or village. Unlike direct public/supra-local management, it does not form part of a technical agency such as the Ministry of Water Supply, the Department of Interior Affairs or a Public Works Department.   Public water PLC's are quite common in Western Europe, where they can be found in, e.g. Germany, the Netherlands, Belgium, and in the United States. Though they are rare in the rest of the world, examples can be found in the Philippines, Chile, and South Africa, among others.

3.1.4   IWRM Requirements at Constitutional Functional Level

Based on foregoing and on the outcome of the three test cases IWRM requires from the constitutional function a system that:

- enables effective development and implementation of laws and regulations,
- enables effective constitution and development of relevant institutions,
- regulates decision making based on interests of all stakeholders,
- enables all stakeholders to participate in decision making,
- provides quantitative and qualitative standards for use,
- provides quantitative and qualitative standards for effluents,
- enables and regulates effective control and sanctioning of violations,
- enables implementing agencies to take the necessary steps to secure and conserve the resource,
- provides effective and transparent accountability mechanisms.
- provides sufficient capable people to meet the IWRM demands of policy making, adapting legislation and all other activities
- enables and regulates private sector participation

3.2 THE ORGANIZATIONAL FUNCTION: INTEGRATED WATER RESOURCES MANAGEMENT

3.2.1 Activities of the Organizational Function

The organizational function is integrated water resources management. The ultimate goal of the management process is to allocate water in quantity and quality terms for different purposes. The process involves resource assessments, planning, decision making, implementation and policing on allocations and use of water resources with and based on the interest of stakeholders. These processes are time and location specific. However, some general activities in this complex process can be identified and are listed below (Savenije, 1997):

Assessment:
- water resources assessment (quality and quantity)
- environmental assessment

Planning and decision making:
- problem analysis
- activity analysis
- demand analysis and demand forecasting
- formulation of objectives and constraints
- design of alternative water resource systems
- system analysis
- system simulation and optimization
- sensitivity analysis
- multi criteria and multi constraint tradeoff analysis
- selection and decision making
- involvement of stakeholders

Implementation
- allocation of water resources
- demand management
- administration of service provision to water institutions
- operation and maintenance
- monitoring and evaluation
- financial management and performance auditing
- communication, negotiation and conflict resolution

Policing
- inspection and control
- sanctioning

The activities are highly multidisciplinary, involving engineers in hydrology, hydraulics, construction, water supply, sanitation, hydropower, irrigation, and non-engineers such as: environmentalists, ecologists, lawyers, economists, sociologists, agriculturists, politicians and representatives of interested parties, pressure groups, and water users.

Table 2   PSP options with their characteristics

PSP option	description	contract duration	benefits	costs	risks
Community Contractors	community based organisation that carries out particular tasks in that area	flexible but usually short term	- control within community - cash remains in community - empowerment - skill enhancing	social development and (continuing) technical support to set up and train	- distortion by local political leaders, - misuse of funds
Mini contractors	small scale contractors or individual artisans	flexible but usually short term	- develop private enterprise skills, - money retained in local area	- training of small contactors - revolving funds for tools and equipment	- high failure rate among starters - inexperienced contactors failing
Suppliers	manufacturers and retailers of materials, equipment and spares	one time to long term renewable	- develop private enterprise skills, - money retained in local area	- excessive cost through high markups; - training of small suppliers required; - revolving funds needed	- high failure rate among starters - inexperienced suppliers failing
Contracting out, Service Contracts	services contracts for managing operating or maintaining part of the assets (pumping stations etc.), provision of administrative services (billing)	one - two years	- develops private enterprise skills, - money retained in wider area - promotes competition, - reduced risk contact failure	- supervision - regular tendering	
Management contracts	The contractor, for a fixed fee, takes responsibility for managing operations and maintenance of existing assets	3 - 5 years	- improved service with reduced risk to client (government), - potential for setting performance standards	- more expensive than using public sector managers, - transaction costs through contract preparation	- difficulties in setting and measuring performance standards; - little incentive for good consumer services

Lease	The contractor takes responsibility for operating and maintaining an existing system and for collecting the tariffs from which the contractor makes its profits and pays its cost. Government remains owner of the fixed assets and is responsible for investments in new works	8-15 years	profitability depends upon increasing efficiency of managing the assets, considerable amount of risks transferred to contractor	-transaction cost in setting up contract; - need for contract monitoring; - requirement for enabling legislation	- unknown condition of assets at start of lease;
Concessions	The contractor takes responsibility for operating and maintaining an existing system and for new investments, paid for by collecting tariffs from which the contractor makes its profits and pays its cost. Government remains owner of the fixed assets.	20-30 years	relieves govt of need to fund investments and management operations	- high transaction cost - significant loss of government control - robust legal framework needed - higher profit margins required as risk increases	- negotiation and flexibility required to a considerable degree - termination of contracts would create considerable problems
BOT/BOOT	Build (own) Operate Transfer contract is used for a complete new segment of a system such as water source development, hydropower stations, treatment works. The contractor is paid for bulk services at a rate which should give him reasonable profits	20-30 years	- mobilises private finance for costly new investments - high level of service (delivery)	- high transaction cost - significant loss of government control - robust legal framework needed - higher profit margins required as risk increases	-weakness in revenue generation depletes utilities ability to pay for BOT bulk service.
Divestiture	the assets and rights to operate and maintain them in exchange for tariffs are sold to a private company or though a sale of shares or to the existing management through a "buy-out"	indefinite	a private company would have clear incentives to respond to changing demands and achieve full cost recovery	high initial cost in selling	like all disadvantages for concessions, - private sector monopoly

3.2.2   Development of Integrated Water Resources Management Capacity and Capability

The development of an integrated water resources management capacity and capability is both a top-down and bottom-up process. The top-down process is a result of the execution of the care function of government. Government has to impose measures and regulations to protect the interest of society through protection of resources, ecosystems and socioeconomic well being. Government executes this task through policy development and creation of legal and institutional frameworks for use and management of water resources.

The bottom up approach originates from the operational level where different and sometimes conflicting use and control interests need to be protected. This bottom up process is to be carried out in the enabling environment as created by government. As this is a process of learning, correcting and adjusting, the frameworks as imposed by the constitutional level should leave enough room for refining and adjusting. This means that only main policies and major concepts are regulated in law and the interest groups are given the opportunity to formulate their own way of coordination and operation. This of course should be done under tutelage of government.

The development efforts should be focussed at the creation of a platform for weighing interests and decision making on water use and control. To be successful this platform should have the support of the stakeholders. A consultation process before establishment of such platform is warranted.

The platform should have a decision making capacity on river basin level that reflects the interests of different uses and users. The great lines for decision making procedures should be part of the regulatory framework prepared by government. This should include a clear regulatory framework with norms and standards for decision making. Details could be worked out by the stakeholders. The agreement on procedures (byelaws) has of course to be approved by government.

An effective and transparent accountability mechanism is essential for effective management. This includes accountability to the operational level to see whether the service agreements are being carried out, and to the constitutional level to ensure that societies' interests (wise use of resources) are not violated. Such accountability mechanisms require the platform to have power to control and sanction violations. These accountability mechanisms and policing powers have to be regulated in legislation from government.

An effective IWRM system requires reliable information on the availability, use and quality of surface and ground water in the basin. Data bases, observation networks and inspection systems are to be made accessible, improved or developed. Good access to these data allows analysis of several options-scenarios for interventions in development and use of water at basin level. Sufficient capable and motivated people with the appropriate tools are required to meet these IWRM demands on planning and management, control and development. Identification and development of people and tools for management are part of the development process of the platform that also requires consent and support from the different stakeholders as important cost can be involved. The legal and institutional framework in which the platform is to be developed and will operate is to be created at constitutional level. At constitutional level also the policies are set for development of capable and motivated staff.

3.2.3   IWRM Requirements at Organizational Functional Level

The basic function at organizational level is to coordinate between the different interest and decide the different uses of water. An effective organizational function requires:

*   a decision making capacity on (sub-) river basin level which reflects the interests of different uses and users,
*   a clear regulatory framework with norms and standards for decision making,
*   a system that provides reliable information on the availability, use and quality of surface and ground water in the (sub-) basin,
*   a system that allows analysis of several scenarios for interventions in use of water at basin level,
*   an effective and transparent accountability mechanism,
*   power to control and sanction violations,
*   sufficient capable people to meet the IWRM demands on planning and management, control and development.

3.3   THE OPERATIONAL FUNCTION: WATER SERVICES

In IWRM a distinction is to be made between the management of the water resource and the delivery of water services, both of which are necessary in each country. Usually the planning, development and management of the water resource must be a government responsibility to ensure that public interests is served. In contrast, specific water services are generally best delivered by autonomous and accountable public, private or cooperative agencies with scope for increased private sector participation.

Manipulation of flows and (ground) water levels to provide these services, requires an hydraulic infrastructure which development and management cost needs to be recovered from the beneficiaries or from the community at large if the system is to be sustainable.

Sustainability requires among others, adequate funding for operation, maintenance and management of the system. The costs for services are in principle to be recovered from those who benefit from the provision of those services. This requires an identification of beneficiaries and clients for the services provided[3]. Clients are only willing to pay for services if these are reliable and considered not too expensive. Often government subsidies are provided to reduce the cost for the clients or to stimulate certain developments. However, this usually reduces the incentive for the managing agency for optimal performance of service delivery and effective and efficient use of resources. Financial autonomy of the managing agency that is fully accountable to the clients is a prerequisite for system sustainability.

The provision of the services requires an infrastructure that needs to be planned, designed, constructed, operated maintained and after some time replaced or modernised. At the onset of

---

[3]   Clients and beneficiaries are not always identical: for example the clients of irrigation services are usually farmers but often small industries or households benefit from the availability of water in the irrigation canals.

development, the infrastructure is designed to provide a certain level of service. The cost of service provision is directly related to the level of service provided. The higher the level of service, the more management efforts and Infra-structural requirements are needed, the higher the cost. In a situation where clients fully pay for the cost, the level of service must then be balanced against the associated cost in a consultative process with the clients and other stakeholders. They will agree on the level of service and its associated cost. The results are included in a service agreement between the service provider and the client. These agreements can only be successfully carried out if transparent and effective accountability mechanisms and accountancy systems are in place (van Hofwegen and Schultz, 1997). These also constitute part of the service agreement.

3.3.1 Water Management Services

Water management is the manipulation of surface or subsurface flows, levels and quality of water to serve either one or a combination of the following purposes:
*   water supply for agriculture, domestic, municipal and industrial use, recreation and environmental protection.
*   drainage of urban and rural areas;
*   flood protection for urban and rural areas;
*   control or maintenance of water quality.

These manipulations are carried out by individuals or organisations in a provision of public or private services on a local, regional and international scale and are mutually interactive.

The nature of water as a resource and its multiple use requires coordinated efforts to manage the different and often conflicting manipulations needed to fulfill the demands for the different purposes. These management efforts are offered as services and can be carried out by one or more institutions that can be either government, semi-government, private or users' organisations.

It happens that different organisations are involved in the provision of one service. Typical for such situations is the provision of irrigation water for agriculture not directly to the individual clients but through a water user's association.

3.3.2 The Clients

In water management different services are provided for different client groups. The nature of the service determines whether the clients are clearly identifiable individuals who can voluntarily use or reject the services without doing harm to others. In water supply the transaction of a certain volume of water can be demanded or rejected without harming another water user (provided water is not a constraining factor).

However, flood protection, drainage and water treatment are water management activities that are of public nature and cannot be accepted or rejected by individuals. In table 1 the clients of different services are listed. From this table services are clearly not always provided to clear identifiable individuals. A clear definition of the clients is necessary to decide with whom to enter into a service agreement, who is to be charged and where to send the bill.

Table 3   Water services, clients and the service interface

Service	Service interface	Client
Domestic, municipal and industrial water supply	Individual connections with meters	house owner - tenants, households, industries and other identifiable entities.
	Public Taps	community or user group
Hydropower	Individual Intake	hydropower company
Irrigation	Individual Off takes	land owner - tenant
	Tertiary Off takes	farmer group or water users association: land owners and tenants
Agricultural Drainage	Individual Outlet	land owner - tenant
	Tertiary Outlet	farmer group or water users association: land owners and tenants
Urban Drainage	Public storm water sewage, no distinctive interface	all individuals inside service area: land owners, property owners and users
Flood Protection	Dikes, no distinctive interface	all individuals inside protected area: land owners, property owners and users
Waste Water Treatment	Sewage connection	all water polluters: households, industries and other commercial or administrative water using entities

3.3.3   Development of the Service Relationships

As described above, the service relations and the services provided need to be clearly defined and transparent administration and effective accountability mechanisms need to be introduced. These are the main ingredients of the service agreements.

The Service Agreement

For all service relationships it is necessary to define services with their conditions and the compensation required for providing and obtaining these services. These can be formulated in service agreements that contain the services delivered, the level of service, the payment for service, the procedures for monitoring and control of fulfilment of obligations, for conflict resolution and for changing the agreements (Snellen, 1996). These agreements can be formal and informal.

Box 2: Service Agreements

Service Agreements consist of:

Transactions

1. Services Provided and the level of service provision
2. Payment arrangements for these Services

Accountability Mechanism

3. Procedures to check whether obligations are met
4. Consequences for not fulfilling the agreement
5. Authority that will be addressed in case of conflict
6. Procedures that will be used for updating and improving the agreement.

Service Agreements (box 2) consist of two main ingredients: Transactions and Accountability Mechanisms. The transactions deal with the provision of the agreed level of service (in water delivery   protection levels, drainage levels, effluent charges) and the payments from the clients (how much, when, where and how). The accountability mechanism is to ensure that obligations from both parties are met and that the provision of services and its payment arrangements can be adjusted according agreed procedures.

Level of Service

The level of service can be defined as a set of operational standards set by the managing agency in consultation with users, other affected parties and the government to manage a water utility (van Hofwegen and Malano, 1997).

A consultative process, which includes government, users and agency is required to establish the agreed level of service, the procedures involved in establishing a level of service and the price that has to be paid for it. The willingness and capability of clients and beneficiaries to pay for the services depends on the profitability of their enterprises and the reliability of services. With a higher level of service the clients reduce their risks and are willing to invest in higher value production (crops, machines etc).

Agreed and Declared Levels of Service

Performance oriented management is related to the achievement of the provision of a predetermined level of service. This level of service is greatly influenced by the physical environment (availability of water and land resources), management environment (institutional and legal frameworks) and the hydraulic infrastructure.

The establishment of the level of service is part of a development process that requires a continuous process of consultation, adjustment and adaptation within an effective accountability system. In this process a differentiation has to be made between an agreed level of service, which is demand driven, and a declared level of service, which is supply driven. The difference relates to the degree of consultation during the process of defining the level of service and the level of cost to be recovered from the beneficiaries and the government subsidies related to that. The services provided can thus be dictated by either the clients or the management agency. Which one can be applied depends on government policies and the enabling environment the government has created.

In the first case we speak of service oriented management based on agreed levels. This is the provision of an <u>agreed level of service at an agreed level of cost</u>. This agreed level of service is the result of a consultation process in which all the interested parties (government, agency, clients) are involved. The level of service to be provided is balanced against the cost associated with the delivery of that particular level of service. The objectives and targets for delivery of services are formulated in a service agreement. In service oriented management the needs of the clients are of primary importance. The service provider adjusts its management practices according the outcome of the consultation process to these needs. The performance of the managing organisation is measured against the fulfilment of the agreed level of service.

In the second case we can speak of performance oriented management based on a <u>declared level of service to be provided at a declared level of cost</u>. This means that the need for responding to the needs of the clients is of second order. The clients adjust their consumption or production practices to the services provided. This is usually the case in situations where water resources are scarce. Often an important part of the cost to provide the services are covered by subsidies from government.

In formulating an agreed level of service also a differentiation has to be made between agreements for existing schemes and for new schemes. With existing schemes the level of service is constrained by the facilities present. In new schemes the facilities can be directly adjusted and adapted to the agreed level of service.

Level of Service Determining Factors

Water supply, hydropower and irrigation and drainage organisations operate under a physical, legislative and socioeconomic setting that defines the environment in which the level of service specifications must be developed. The formulation of level of service specifications is a process of 'customisation' of the supply parameters duration, flow rate and frequency in combination with the quality parameters. Several factors condition the customisation process becoming a set of boundary conditions to the final service specifications. These include water rights, water requirements, production or consumption practices, land ownership and size of land holdings, water availability, irrigation policies, water rights and competing water uses.

Level of Service Specifications

The delivery of services is based on a set of specifications that will govern the management of the infrastructure. These specifications serve two purposes: (a) to provide a set of rules against which the operational performance of the system can be measured, and (b) provide a set of rules that govern the delivery of service. The first set of specifications are clear, quantifiable and measurable operational parameters that permit the comparison of the actual operation against the management targets. The second set of rules refer in general to the conditions under which the services are provided. These specifications and conditions are part of the Service Agreement between the service providers and their clients.

Quality Criteria for Water Delivery Service

The quality or level of service is related to the form of water delivery and can be characterised by a combination of parameters. These are adequacy, flexibility, reliability, equity and equality. Furthermore, service quality can also be characterised by the convenience of water delivery and the quality of the water delivered.

3.3.4   Accountability Mechanisms

Accountability is the extent to which the performance of and organization and its staff can be monitored and controlled either by senior officials, investors or clients. Accountability requires transparency. An organization is transparent if any stakeholder can easily find information about activities and performance of the organization, such as:

- what it is doing
- what results it is achieving
- what are its plans for the future?
- who takes decisions?
- what are the reasons for decisions?

An organization is accountable if it is clear
- what it is aiming to do
- what levels of service it is promising to provide to its clients
- what targets it sets for itself?
- it takes positive corrective actions, after any failures or errors or service deficiencies that may occur

An organization is accountable to its users if any dissatisfied user of its services knows a clear path for making his complaint, and also feels confident of getting compensation for the failure or mistake as the services are well defined and failure of delivery and the consequences are regulated in service agreements.

Accountability is an important feature of any management model. There are at least three different areas in which information from assessments can be applied in judging accountability (Small and Svendsen 1990): the internal processes of the organization managing the service system (service provider), the relationship between the service provider and its supervising board or body, and the relationship between the clients of the system and the service provider.

Different concepts of agency-client accountability mechanisms can be distinguished by level of management and directness of interaction. Three levels of client-agency accountability exist (van Hofwegen, 1996) in parallel to the levels of working rules as formulated by Ostrom (1992)[4].

The first level is of *operational accountability* being the mechanisms for monitoring, evaluation, control and sanctioning of the implementation of the service agreement. This includes daily monitoring and control (i) by the clients of actual delivery of services and (ii) by the agency on client contributions payments for services delivered.

The second level is the *strategic accountability* being the mechanism for monitoring and control by clients on the content of the service agreement e.g. planning process of water allocation, control and division. Water users or their representatives participate in this process actively or they approve proposals prepared by the agency. The result of this process will be part of a service agreement.

The third level is the *constitutional accountability* being the mechanism for monitoring and control by the clients on procedures for strategic decision making of the agency.

A second concept for looking at accountability mechanisms is the degree of directness of

---

[4]   Ostrom identifies three levels of working rules: operational rules, collective choice rules and constitutional rules.

accountability. In direct accountability systems the management is directly accountable to the clients through for example a board of beneficiaries overseeing and approving the work of the agency's executives. In indirect accountability systems the managers are either (i) accountable to a board in which elected representatives of the beneficiaries and stakeholders are taking part, or (ii) to a board which consists of representatives of stakeholders or agencies involved in which no beneficiaries are taking part, or (iii) to their superiors in ministries and departments who are influenced by political parties and pressure groups.

All accountability systems require their own sets of control and sanctioning mechanisms and specific performance indicators and standards. The accompanying accountability mechanisms are often formal and according to laws and regulations. However, often non-official accountability mechanisms exist (rules in use) that influence the performance of the schemes in both positive and negative ways.

3.3.5 Development of the Capacity and Capability of the Service Provider

The change into service oriented management requires a considerable readjustment of agencies operations. The interventions are of technical, financial, administrative and management nature and comprise:
- improvement or modernization of the *hydraulic infrastructure* (intakes, treatment plants, alignments and capacities of conveyance and distribution networks, flow control systems, measuring devices etc.),
- a change in *administration and management procedures* related to a shift from public administration to private corporate management, and
- a change in *operation and maintenance procedures.*

Improvement of hydraulic infrastructure is often necessary to allow provision of the agreed services in a reliable manner and to reduce operational cost. Special attention needs to be paid to the hydraulic aspects of accountability: measuring devices at the interface of service delivery which can be used, measured and understood by both parties.

The introduction of effective accountability mechanisms requires different attitudes in management and demands transparent decision making and administration. Cost administration and budget allocation need to be in line with the needs of the clients. If the price of service is related to the cost for provision of this service, analytic administration is required which allocates the actual cost to those services the cost are made for. Financial management usually has to adjust to this new system. Also a change is to be made from supply oriented budgeting to needs based budgeting. Managing agencies need to have the right to develop financial reserves to allow for a more flexible use of financial resources.

The process of operation and maintenance will need some modification as well. Operational decisions need to be made on clearly formulated rules which are understood by both the operators and the users. Systems have to be developed which register the actual delivery of services for accountability reasons. These systems might require adjustment and streamlining of operational activities in planning and implementing water distribution. Transparency is required in decision making in water allocation and distribution, especially when water is scarce. This requires an administration and publication of conditions and situations which necessitated the decisions taken.

Annual reports on financial and technical performance are a good means.

The institutions and all the individuals within them require incentives to motivate them to behave in a direction towards the common goals. Staff needs to be adequately trained but also needs to be motivated to implement their job properly. Career opportunities and proper payment are prerequisites for performing organisations. Staff development (but also often staff reductions!) are an integral part of any management strategy.

Effective and efficient use of water can only be established if there is a relationship of confidence and trust between the service providers and their clients. Development of such a relationship requires investments in money and time through programmes concerning awareness creation (active and passive information of clients), training and support for introduction of new technology and management systems etc.

3.3.6   IWRM Requirements at operational functional level

Effective operational functioning within an IWRM context requires a management system that responds to societal needs. This means that for water services the system should enable, provide or regulate:

- effective control of the service providers by users/clients and the IWRM Platform
- representation of clients interests at and by the managing agency
- cost recovery by the service provider.
- negotiations between the managing agency/service provider and its clients on the level of service it provides and recovery of its associated cost,
- assessment of the demands, actual use and availability of water (quantitative and qualitative)
- power at the service provider to control and sanction violations
- sufficient capable people to meet the IWRM demands, planning, development and management of services provided
- a system that allows market incentives to make most economic use of water through participation of private sector.

In order to arrive at this set of requirements a process of capacity building for IWRM is required in a service oriented environment. These processes will be described in the following chapters.

CHAPTER 4

Interventions or IWRM Capacity Building

4.1   THE CONCEPT OF CAPACITY BUILDING

Capacity building basically is the development of a management capability and capacity to achieve (sector) formulated objectives and to adjust to a changed management environment. The objective of capacity building in the water sector is to improve the quality of decision making, sector efficiency and managerial performance in planning and implementation of sector programmes and projects to achieve efficient and effective management of water resources and delivery of water services. More specifically, capacity building for sustainable water resources management is designed to improve the capabilities of assessing water resources, facilitate better planning in the context of development planning and promote financially and environmentally sustainable, more efficient and more effective delivery of water services.

Capacity building involves human resources development, the development of organisations and the emergence of an overall policy environment conducive to development. The framework for national capacity building is provided by national systems of planning and management, and their translation into policies and programmes. Capacity building in water resources management is defined as follows:

> *Water Sector Capacity Building is the development of favourable policy environments and institutions, needed for sustainable water resources utilization with respect to all stakeholders.*

Water Sector Capacity Building consists of (Alaerts et. al. 1991):
- the creation of an enabling environment with appropriate policy and legal frameworks; policy issues to be addressed include a focus on sustainable development, water as an economic good, and the principle of cost recovery; and
- institutional development, preferably building on existing institutions; institutional development includes national, local, governmental, public and private institutions, and community participation; and
- human resources development (which includes education, training, career development and performance incentives) and strengthening of managerial systems at all levels.

4.2   THE PROCESS

The capacity building process is initiated by a Government with a firm commitment to improving water resources management by capacity building at all levels. Capacity building puts institutional development and strengthening and other training interventions up-front; as a   result the institutions operate better and are better able to gradually provide products and services in function of the developing demand.

Capacity building for sustainable water sector development not only aims at 'vertical' capacity building within the sub-sectors of irrigation, water supply and sanitation, health or environment, but also seeks to reinforce operational 'horizontal' linkages between these sub-sectors. The process typically consists of a series of subsequent steps.

Firstly, a clear understanding should exist about the water resources and their physical environment, the state and structure of the water sector, and its role in national development plans and in the national concept of environmental sustainability. This analysis is to be as comprehensive as possible, and has to take a long-term view. Importantly, the assessment in itself proves to be a powerful incentive to commence mobilizing staff, resources and political support for sector enhancement. The process offers an opportunity for joint problem definition and strategy development, in which the sub-sectors have to cooperate.

Secondly, as a result a longer-term strategy (describing options and scenarios) and a short-term action plan are to be outlined. If the sector proves very resilient and well developed, capacity building can be restricted to the required activities to maintain this position and timely inclusion of new developments. In many countries, however, the sector may show many structural deficiencies. In such cases the assessment should be designed to identify priority actions that concentrate on interventions that are most likely to have a maximum and broad effect. Being faced by too many weaknesses often proves discouraging, and it is not feasible to attempt to remedy all weaknesses at the same time. Selecting a small number of key issues usually helps to keep the process manageable, achieve success, and start changing attitudes and procedures at large. Common key weaknesses pertain to

- institutional and economic arrangements that distort economic value and price of water and other environmental services;
- lack of integration of the sub-sectors; and
- poor legal and institutional arrangements that impede delegation of responsibility to lower administrative levels, including rural and peri-urban communities and non-governmental organizations.

Typically, an analysis will first describe the perceived technical or physical problem. Then the institutional cause for this problem is to be identified and specified. Thereafter it becomes clear which capacity building instruments could be applied to address the institutional problem.

4.3   MODELS FOR CAPACITY BUILDING

Once the focus for the capacity building is decided upon, the capacity building can be organized according to a particular coherent *model* that is composed of *instruments*, or interventions. As

capacity building is a long-term and country-specific process, the experience with comprehensive successful models is still limited. Comprehensive water sector exercises are rare, and typical models deal comprehensively with key issues in the sector, including:
- retaining first order interests as the responsibility of government
- decentralization and privatization of activities related to second order interests
- transfer of management authority to users
- autonomy of management organizations
- internalizing the scarcity value of water and deregulating water prices.
- strengthening local capacity builders
- education and training
- changing managerial systems and the organization's culture.

None of these models can be implemented successfully if undertaken as a single action. Changes in the legal, administrative or economic frameworks have consequences reaching deep into many other institutions. New policies can only be implemented well if the capacity of all relevant actors is built simultaneously.

4.4 INSTRUMENTS FOR CAPACITY BUILDING AND HUMAN RESOURCES DEVELOPMENT

The instruments that are applied in these models typically encompass:
- reform and adjustments in legal and regulatory frameworks.
- reform and adjustments in administrative procedures.
- local and international financial and technical assistance.
- workshops and seminars to exchange knowledge and experiences, or stimulate joint problem identification and problem solving.
- education and training, to acquire new insights, skills and attitudes with respect to technology, managerial or behavioral sciences, or integrated approaches.
- data base and research capacity development, and development of management information systems (MIS).
- networking and development of information and communication channels and mechanisms.
- twinning arrangements between peer organizations from different continents, networking between peers to learn from each other's experiences, and relationships with professional associations.
- creation of platforms (meeting occasions, secretariats, procedures) at international level under multi-lateral auspices to prepare for international cooperation notably in water sharing.
- hardware: training materials, libraries, journals and newsletters, computers, data-processing and -storage equipment, communication equipment, etc.

4.5 HUMAN RESOURCES DEVELOPMENT - EDUCATION AND TRAINING

Introduction and application of IWRM is needed to manage water resources and improve water services that will sustain human and economic development. Government should provide leadership, commitment and a focus on principles to direct this process. Effective capacity building, requires a good cadre of capable senior managers. To obtain such cadre, long and short term plans should be made for education and training.

Education is an investment in the future and aims at transferring knowledge, insight, methodologies, skills and, importantly, new attitudes. Training aims at more specific problems, implies shorter contact time and attempts to offer only directly applicable skills that are required for a given setting. The aim of sector related education and training is to achieve performance objectives by fulfilling the requirements on knowledge and skills to develop and implement sector programmes. Moreover it should provide an adequate understanding of processes related to the sector programmes to allow formulation and implementation of corrective actions.

A systematic human resource development strategy has specific objectives:
* to ensure that management, technical and administrative skills necessary to fulfil the charter and objectives of the organisation are developed and retained within the firm or department;
* to optimize the opportunity for personal development and work satisfaction among individual staff members.

Such programs provide all employees in the enterprise or department with access to appropriate training at crucial times during their service. The training activities are designed to build up the technical and managerial proficiency of the organisation, make it more flexible and adaptable and strengthen the commitment of its staff.

Training is an important tool of management and should as such be an integral part of any management strategy. In such management strategy a balance is made between investments in infrastructure and procedures on the one hand and the training (human resource development) needs on the other hand. Changing management environments, new technological developments and growing awareness to use water (cost) effectively and efficiently will put a continuous demand on training and development of agency staff and consequently on the development of a training capability.

A continuing commitment to systematic irrigation training is absolutely essential. This requires the inclusion of development and management of an internal or external training capacity in the training strategies. A well-defined, systematic and comprehensive training strategy can form a good basis for demonstrating to domestic and external funding agencies the needs for support for training and the assurance that additional investments in training will lead to improved water resources management.

Provision of Education and Training Services

Providers of education and training services can be categorized along lines of the status of the organization, the level and type of education and training services and the (inter)national orientation. The status of the organization providing the education and training services can be external (public or private) or internal (employer), national or international. Most publicly financed or provided training occurs before employment. Private institutes can meet an important share of the additional required skills without public financing if they are allowed to compete freely. However, private training is often too tightly regulated. Ceilings on tuition constrain income, which prevents entry into occupational areas with high equipment and instructor costs. Regulations requiring private institutes to follow the public curriculum also reduce their flexibility. Employer training is financed, organized and managed by the employer to fill up the deficits in skills and knowledge required for the specific tasks in the organization. The training focuses at the requirements of the employer and is usually part of a career development plan.

When capacity for under- or post-graduate education needs to be strengthened public educational institutes may be insufficiently specialized and too oriented to academic research agendas that bear little relevance to local problems. Often academic institutes need support to reform their curricula to (i) focus better on local technical and multi-disciplinary problems, (ii) introduce more interactive and stimulating teaching methodologies, such as integrated multi-disciplinary group works, field work and workshops, (iii) involve practitioners in the teaching program, and (iv) shift from teaching factual knowledge to developing skills and attitudes, emphasizing adaptive design and structured learning approach.

Research institutes have an important role in education and training as they provide information on actual issues and compilations of methodologies to identify, analyze and resolve problems related to development and management land and water systems in the context of agenda 21. These research institutes can very well be part of managing agencies or education and training institutes.

CHAPTER 5

Assessment of the Institutional Framework - Process and Tools

As explained in chapter 1 the analytical framework for the assessment of the institutional framework is using the process to come from an identified present water resources management situation to a desired integrated water resources management situation. The steps in this process are identification of the present situation, formulation of a desired IWRM situation, formulation of interventions to arrive at the desired IWRM situation and establishment of a monitoring system to see whether the interventions are being carried out properly and whether they really contribute to the achievement of the IWRM goals.

The assessment process is using ten steps which have been based on the experience gained from the test cases in Guatemala, Colombia and Jamaica. In this chapter these steps are elaborated and tools which have been used successfully in these case studies are presented.

5.1   STEP 1: EXISTING WATER MANAGEMENT SITUATION

The present situation on water resources use and management should be well known before any intervention directing to IWRM can be made. Understanding the water situation is a prerequisite for assessment and analysis of the institutional framework and the (potential) water use conflicts between stakeholders. It appeared essential to have a basic document on the present water management situation to start the institutional assessment process. Such document will represent an experts opinion and will not necessarily be complete, accurate and representing the opinions, desires and aspirations of all stakeholders.

Important aspects to be dealt with are: water availability and water use, stakeholders, physical conditions, socio economic conditions, legal framework, institutional framework, policies and trends and the financial situation. IDB will assign experts to prepare such (desk study) report describing the existing water management situation combined with registered problems (quantity, quality and environment). The report serves as a general background document for the following steps and has to be disseminated accordingly.

Physical Conditions

The assessment of the physical conditions concentrates on the temporal and spacial availability and use of water (quantitative and qualitative). It requires information on the climate and meteorology, hydro(geo)logy, aquatic eco-systems, abstractions and influents and the availability and capacity of storage facilities.

As in IWRM water is managed on basin or sub-basin level, use of water resources, water distribution per sector and the resulting water balance have to be identified per (sub-)basin. This is essential information for IWRM so the existence of observation networks and data bases, the levels of processing and the accessibility to these data sets and should be included in the inventory. For the inventory a clear distinction has to be made between the different levels : National, Basin

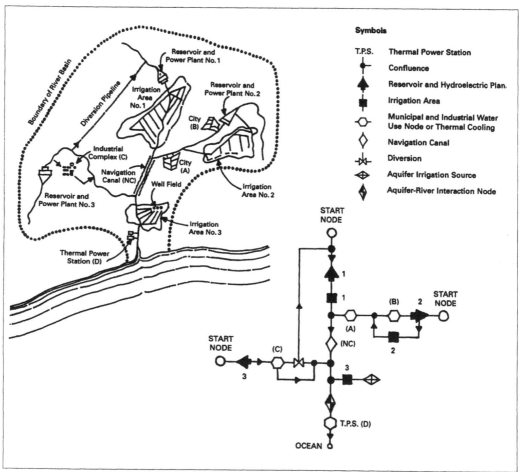

Figure 5 Water use flow diagram in a (sub) basin

and sub-basin level. At national level the inventory should limit itself to the water balance in the different basins. Such water balance provides insight in whether and when the basin is in a surplus and deficit condition. Temporal surplus conditions provides the opportunity to overcome temporal deficit conditions by creation of storage facilities. During deficit conditions the occurrence of major conflicts in interest will be most prominent both in quantitative and qualitative sense.

At basin or sub-basin levels a more detailed inventory can be required. For such an inventory a water use flow diagram as presented in figure 5 can be most helpful. Such a diagram provides not only the uses and users but also their inter-dependencies with regard to water quantitative and qualitative aspects.

Stakeholders and Interest Groups

Stakeholders are people or groups of people with a legitimate interest. Legitimate interests are formulated in the bye-laws of the interest group where the stakeholder is regarded as a private entity/body. Stakeholders are not the same as interest groups. Interest groups represent all kind of interests: public, private, environmental, social etc. If they are organized and have statutes or by-laws they represent legitimate interests (GO's, NGO's, professional organizations, commercial organizations, users associations) and as such become stakeholders.

In IWRM the stakeholders can be classified as follows:
- water users   consumptive and non-consumptive uses: agriculture, domestic water supply, hydropower, tourism, shipping/navigation
- water polluters   agriculture, industry, domestic etc.
- water managers   organizational and operational level
- water policy and law makers   constitutional level
- society   general interests represented by government agencies (e.g. national, city or district councils) and specific interests represented by NGO's (action and pressure groups, environmental organizations etc.).

It depends on the socio-economic and the political situation whether all the interests are represented. So important to assess is which are the stakeholders who's interests are considered and which interests are not considered but are important for sustainability.

The water use flow diagram can be most helpful in identifying the stakeholders. Water use will be different for each basin. Therefore, stakeholders have to be identified on basin level.

Inventory water problems:

In this stage the inventory of water problems limits itself to those generally known and registered at the main stakeholders. The basin water balances and the water use flow diagram can again be most useful to put the registered problems into the basin perspective. The type of problems concern water quantity, quality and environmental problems (erosion, siltation salinity etc.) but also relate to navigation, recreation and other uses. This inventory will be used in the second step as a starting point for analysis of the problems and identification of other interest groups and stakeholders.

Water Rights and water allocation

In most of the countries water is considered a public good, but individuals can obtain private rights over water by tradition. Existing water rights are often a main constraint and a source of many problems in the optimization and introduction of IWRM. The system of water rights (surface and ground water) their acquisition and conditionalities, their transferability and the system of water right administration should be clearly presented. Especially in water market systems a sound administration and a system of approval of transfers is required. If not available, planning of water will become very difficult.

Water allocation between different uses and users is an organizational function. The introduction and development of IWRM could learn much from the present water allocation system, the conditions and procedures and actors in the decision making process. A good description will therefore be very valuable especially for the analysis of water quantity related conflicts or problems. These aspects can be verified in the stakeholders interviews in the next step.

Description of socio-economic and financial environment:

Many of the above identified problems will be said to be due to the actual financial and social situation. Lack of infrastructure and its maintenance, lack of good management and the possibility to have effective cost recovery are usually blamed to lack of financial resources. Therefore it is important to have insight in the budget allocation mechanisms, budget constraints, cost recovery mechanisms, subsidies, price and tariff structures, collection mechanisms, collection efficiencies, capability and willingness to pay for the various uses. These mechanisms should be identified in general terms per sub-sector or use.

Existing water policies and strategies

In many countries the water sector is under debate due to problem experienced and the commitment of governments to the outcome of the international conferences. Though often not yet formulated, many countries are in the process of policy development. These policies deal with principles like: equitable and socially acceptable water distribution (priorities, redistribution to marginal groups: poor, women etc.), water as a scarce, finite and economic good (efficient water use, cost recovery, pricing mechanisms and tariff structures, transferability of water rights, rate of commercialization), water management at the lowest appropriate level and on hydrological boundaries (delegation and decentralization, water users participation, involvement, water management by and for water users), integrated planning arrangements and other coordination efforts, private sector participation, and environmental protection. An inventory of these policies indicates the level of awareness and commitment at constitutional functional level.
Lack of these means either that there are no problems, lack of awareness or lack of political will.

Legal framework

An inventory with an explanation of principles is required of existing water laws (and other relevant environmental legislation), water regulations and relevant environmental regulations and decrees and bye-laws of water authorities and river catchment agencies. Important is to indicate, whether and how the above mentioned focal policy aspects are incorporated in the legislation such as equitable water distribution, pricing, delegation and decentralization, participation, integrated planning, environmental protection.

Especially in countries where water policies are changing, legislation will be under reconsideration or in the process of change. Therefore it is necessary not only to present the existing legislation but also the adjustments envisaged. When legislation is in a process of change this indicates that on constitutional level they are aware that present legislation does not satisfy the needs. It is of great value to describe the background of these changes and the direction of the change.

Special attention is to be paid to "trial" legislation where government has given mandate to certain management entities to work with legislation under design in pilot areas for example IWRM or basin level management.

Relevant water institutions

Relevant water institutions are those institutions that with regard to water and water management either formulate policies and laws, do or are involved in water planning, coordinate water uses and users, provide water services or make use of water services. These can be government, semi-government or private institutions on national, basin or use level. A water use flow diagram can give an indication of water users, service providers and coordinators.

Past and Present experience in IWRM
It is important to know what has been tried in the past to overcome certain problems encountered and to what extend these interventions have been successful or not. A description is desired of lessons learned from local experiences of earlier and/or present interventions in the field of integrated water resources management and reasons for success or failure.

5.2   STEP 2: STAKEHOLDER SELECTION

A first inventory of stakeholders will be made in step one. These stakeholders will be the obvious operators of water services, coordination bodies and policy and law makers. For the further process a selection of stakeholders has to be made to avoid duplication. Also some stakeholders might have been overlooked in the first study. Therefore, under auspices of the IDB a representative but independent team is formed to identify and select relevant stakeholders from the categories: water policy makers, water managers, water service providers, water using agencies, water using groups, water users and other potential interest holders at constitutional, organizational and operational functional levels. These stakeholders will be approached for in depth interviews.

5.3   STEP 3: STAKEHOLDER INTERVIEWS

Experts carry out an elaborate procedure of interviewing the selected stakeholders applying the guidelines for interviews (appendix I). These guidelines are in the format of a questionnaire which contain questions relating to the stakeholder interviewed and its perception of the existing situation and what they consider to be the desired IWRM situation. During this interview previously overlooked stakeholders can be identified through the identification of parties which negatively influence the implementation of the stakeholder's duties.

A different set of questions under the issues in the matrix is used for all three functional levels.

they are organized under the headings:
- stakeholders
- awareness
- policy
- legal framework
- institutional framework
- financial arrangements
- human resources development
- management information systems and decision support systems.

The selected stakeholders will be invited to answer the questions during the interview. It is important to use these questionnaires as a guideline for interviews and not as a questionnaire just to completed. The interviews should provides information on the situation of water management and indicate the conflicts and the level of agreement and disagreement between the stakeholders. It is therefore important that the interviews are made by specialists that understand the meaning, purpose and operationalization of IWRM and the potential problems and conflicts that might be encountered.

A second part of the interview is to obtain a description of the stakeholder's concept for improvement of existing water resources situation country towards a more integrated water resources management. The following aspects and principles should be included:
- Equitable and socially acceptable water distribution
- Efficient and economically sustainable water use
- Delegation, decentralization and other devolution of authority
- Participation of stakeholders
- Integrated planning
- Private sector participation
- Environmental protection.

It is obvious that no guidelines can be prepared on how the IWRM situation should be as this is location and time specific. However, to give direction to the process on formulation of a desired IWRM situation, an "ideal" IWRM situation is formulated where in relation to all the points raised during the interview, a clarification is given on how the situation would look like under ideal conditions.

The formulation of the desired situation by the individual stakeholders provides information of what they consider the main constraints and what should be changed and what they see as being realistic and attainable in their present situation.

The "ideal" IWRM situation is derived from the theory on IWRM and the internationally accepted and applied principles on water policies. Use has been made of regional and local water policy documents aiming at IWRM. (IDB 1997, Worldbank 1993, ADB 1996, South Africa 1997, the Netherlands 1997). The "ideal" IWRM situation does not exist and local and regional conditions will determine what the most appropriate situation will or should be. The ideal situation is only presented to provide an orientation in formulation of the desired IWRM situation. Conscious choices must be made to deviate from the ideal situation. This not only helps to increase understanding of the implications of IWRM but it also generates a better sense of participation and belonging of the end product as it allows for active contribution to the formulation process as the desirable IWRM situation will be used to define gaps in different arrangements which in turn could be used for formulation of interventions (steps 4 - 6).

5.4   STEP 4: ANALYSIS STAKEHOLDER OPINIONS

The guidelines are presented in the format of a matrix where through sets of questions for different stakeholders the present and desired situation for each of these stakeholders on the various levels are identified (figure 6). They outcome of all the interviews will be collected and an inventory will be made of agreements and disagreements between the different stakeholders on the present situation, the problems and constraints and the steps to be taken to come to a better water management. The results of the interviews are described in a report and disseminated with the background document to the interviewed stakeholders. These stakeholders should also be invited to the workshops which follow in the process.

5.5   STEP 5: WORKSHOP 1: PROBLEM IDENTIFICATION

The first workshop to which all the relevant stakeholders are invited deals with the assessment of the existing water resources management situation and problem identification according to the perception of the stakeholders. The steps 1-4 were focussing on individual stakeholders and their interests. Their agreements and disagreements as formulated in the analysis report in step 4 are an interpretation of the "expert". Therefore it is important that all the relevant stakeholders recognize their problems and those of others.

The purpose of the first workshop is to confront the different stakeholders with the perception of other stakeholders and to obtain consensus between all different stakeholders of what the real problems are and which should be addressed. The analysis report will be used as a reference and will be improved in accordance with the outcome of the workshop. The agreed set of problems will then be used as an input for the further stages on formulation of a desired IWRM situation and necessary interventions. During the test cases it proved to be a very fruitful method to arrive at a set of most important problems.

Important is that the workshop will be organized under the auspices of IDB as participants will only take such an activity serious if the result will contribute to the process of change. This means that the outcome should be included in the policy development process, implementation process or operationalization of water resource management. it should be mentioned that the Bank normally does not finance travel and per diem expenses for the workshop participants.

Level:	....				
Stakeholder:	....				

Issues	Present Situation	"ideal" IWRM situation	desired IWRM situation	Gap	Interventions
Awareness					
Policy					
Legal Framework					
Institutional Framework					
Financial Arrangements					
Human Resources Development					
Management Information and Decision Support Systems					

Figure 6 Supporting matrix for assessment of institutional setting and identification of interventions for IWRM

5.6 STEP 6: WORKSHOP 2: FORMULATION OF DESIRED IWRM SITUATION AND INTERVENTIONS

The second workshop (one to three months after the first workshop) will elaborate extensively on the principles of integrated water resources management and will further result in the formulation of a desired water resources management situation in that specific country or river (sub-) basin and the set of interventions that will be needed to achieve that.

This workshop is indicative and the outcome provides directions for constitutional, organizational and operational interventions. The outcome should be seen as an input for national policy and decision makers on the one hand and as a framework for defining interventions at the three levels for IDB activities. It is therefore important that the status of the outcome is valued in this light.

In case these guidelines are applied for specific project work, an additional step in this workshop is required to analyse which, if any, of the above formulated interventions should be promoted in the context of this specific project and which interventions better to leave for other projects or sets of activities.

5.7   STEP 7: PRELIMINARY COUNTRY/BASIN/SUB-BASIN REPORT

Based on the foregoing steps the experts will draft a preliminary country document comprising:
- assessment of existing water management situation
- complete problem inventory
- desired water resources management situation
- proposed set of general interventions needed to reach the desired situation and, if feasible, suggested specific interventions to be included in Bank operations

5.8   STEP 8: DISSEMINATION AND COMMENTS.

The draft country/basin/sub-basin report is disseminated and a thorough procedure for collecting comments from the different stakeholders at the different levels is followed.

5.9   STEP 9: FINAL COUNTRY/BASIN/SUB-BASIN REPORT

Experts draft a final country/basin/sub-basin report which is offered to the IDB for endorsement and inclusion into the strategy and/or into specific water related projects for the specific country.

5.10   STEP 10: MONITORING PROCEDURE

The IDB designs a monitoring procedure to follow whether the interventions are taking place and whether the envisaged results are achieved.

CHAPTER 6

Application of Guidelines and Lessons Learned

These guidelines have been developed in an interactive process which included field tests in three countries (Colombia, Guatemala and Jamaica) and a desk study in one country (Chile). The objective of the field-tests was to use, test and improve the analytical framework and the guidelines for its application. During the test local experts carried out the first two steps in the framework cycle: a description of the existing water resources management situation and an assessment of the desired IWRM situation. This was done by preparing an inventory of the water management situation based on available data and knowledge followed by interviews with relevant stakeholders. Their findings were documented in "country reports" which served as a basic input for a workshop with relevant stakeholders to discuss framework, outcome and possible measures towards integrating water resources management. As the aim was testing the methodology, the IWRM content of the outcome can not be taken as representative for the respective countries. The results of the process tests lead to an improvement of the analytical framework, guidelines and interview procedures.

Application of the guidelines in the project cycle

The use of the guidelines initiates a process towards balancing the interests of different stakeholders in water. The guidelines can be applied in different stages of the project cycle: sector policy making, sector planning, institutional design and management arrangements for the sector and for specific projects. Its use should result in an agreement on what the problems and conflicts are and how these can be resolved. The process in itself is cyclic and by monitoring the effectiveness of the interventions new problems and constraints can be identified and corrective actions or new solutions have to be sought. The following points require attention in the application of the guidelines:

1.   The use of the guidelines by IDB teams has to be regarded as the initiation of the process towards IWRM attached to projects envisaged. The first cycle of the process results in a set of interventions necessary to achieve the desired IWRM situation. These interventions can be included in the project design. These projects can be sectoral on regional or (sub-) basin scale or sub-sectoral and on local scale.

2.   The application of the framework is most effective in programs aiming at sector wide institutional change and development because the aim of the program coincides with the

purpose of the guidelines. Moreover, the program is most likely supported by the main stakeholders making the possibilities for interventions wider and necessary adjustments in legal and institutional frameworks less complicated.

The guidelines can also be applied in relation to local projects in physical infrastructure. The project should be of such scale that different stakeholders will be influenced and conflicts of interests on local and (sub-)basin scale are foreseen. However, in this case the focus of the project is on the physical works and institutional change is a derivative of such project. The possibilities for interventions will also be limited by the room provided in legislation as it can hardly be expected that for only one such project legislation will be amended.

3   It is important to notice that on local scale the situation is not always perceived as problematic. However, it is the duty of government to foresee possible negative effects for and conflicts with the interest of society. In such case government should take appropriate action through awareness creation and in a later stage, should participate as a stakeholder in the formulation of interventions using the proposed framework.

4.   The guidelines can be applied at different levels of scale: sub-basin, basin and national level. For whatever level of scale these guidelines are applied, it is crucial to identify and engage all relevant stakeholders at the three functional levels. Leaving out some stakeholders might lead to non-acceptance of the outcome of the process and obstruction of the further development of IWRM. Therefore, it is imperative that these exercises cannot be done through desk studies.

5.   It is advisable to engage independent local experts and preferably not from within the government. Independency should take away bias towards selection of stakeholders in the process. Government officials are likely to focus on official government policies and government agencies limiting the margins of problem identification and solving.

6.   During the tests it became clear that the interview procedure required more emphasis. The purpose of the interviews is to obtain the opinion of the individual stakeholders or their representatives. The guidelines for the interviews are meant to be a tool for the interviewer to structure the interview and to interact with the stakeholder on the different issues raised. The guidelines should not be handled as questionnaires to be handed over to the stakeholder to be filled as then the sensitive issues will not surface. This means that besides a good understanding of IWRM, good communication skills are required for the interviewer. It also emphasises the necessity for good, clear and field tested guidelines.

7.   The workshops proved to be an effective tool to obtain consensus on what the problems and conflicts are and what steps should be taken to resolve them. On several occasions it seemed to be the first time that different stakeholders were sitting in one room discussing their problems! However, to deal with problem identification and resolution in one workshop was too much asked for. Several workshops with different focus are needed e.g. one workshop on problem identification and one on solutions and interventions.

8. Clear prospects for the stakeholders are a necessary condition for the stakeholders to participate actively. The idea that problems are being inventoried and ways are sought to solve them leads to expectations that follow up will be given. Therefore the framework of this process must be made clear from the onset. Active participation also depends on the authority of the initiator of the process. The role of IDB as initiator and organizer of the workshops appeared to be crucial in all the test cases.

9. Some understanding of IWRM and private sector participation (PSP) among the participants in the process is a condition for a good outcome. The first inventory should identify the level of awareness and knowledge on IWRM and PSP. If necessary awareness and knowledge can be raised through information and education and training programmes.

10. A basic requirement for IWRM is the preparedness to reflect on principles of active democracy because IWRM is about weighing private and public interests and therefore a matter of compromises. Outcome of the democratic processes should be respected and solutions should not be forced.

11. The use of the guidelines involves use of time, financial and human resources. The amounts required depends on the scale and level of the project to which the process is attached, the awareness on IWRM among the stakeholders, the political desire to introduce IWRM, the availability of information and the size and accessibility of the area envisaged. The main cost items are local and international specialists time, travel and living expenses specialists, organisation of workshops, travel and living expenses participants workshops. The Bank normally cannot cover costs related to travel and living expenses for the participants at the workshop.

12. In most cases the scale of the process will not be sufficiently known. In such case step one of the process (inventory) can be separated from the remainder. The outcome of step one should then include a cost estimate for one cycle of the process.

References

ADB (1996)
Towards Effective Water Policy in the Asian and Pacific Region, Asian Development Bank 1996
Ait Kadi M. (1997)
High Water Stress - Low Coping Capability, Morocco's Example, in Mar Del Plata 20 Year Anniversary Seminar, Stockholm International Water Institute pp. 113-126.
Alaerts G.J., T.L.Blair, F.J.A. Hartvelt (1991)
A strategy for Water Sector Capacity Building, Proceedings of the UNDP Symposium, Delft 3-5 June 1991; IHE Report Series no. 24. IHE-Delft
Braadbaart O. and M. Blokland (1998)
Public Water PLC's for Low Income Countries, Water Sector and Utility management Group, IHE-Delft.
Biswas A.K. (1996)
Capacity building for water management: some personal thoughts, Water Resources Development Vol.12/4 December 1996.
EUREAU (1992)
Management Systems of Drinking Water Production and Distribution Services in the EC Memeber States in 1992, EU, Brussels
Falkenmark (1995)
Coping with Water Scarcity under Rapid Population Growth, Conference of SADC Ministers, Pretoria November 23-24, 1995.
Franceys R. (1997)
Private Sector Participation in the Water and Sanitation Sector, DFID Occasional Paper no 3, WEDC Loughborough University, IHE Delft.
Hofwegen, P.J.M. van (1996)
Accountability mechanisms and user participation in three agency managed systems, Transactions 16th ICID Congress Vol I-B Question 46.2. Cairo Egypt
Hofwegen P.J.M. van (1996)
Inventory of Training Strategies and Training Capability for Irrigation Management, Working Document ICID Taskgroup on Training Initiatives.
Hofwegen P.J.M. van, H.M. Malano (1997)
Hydraulic Infrastructure under Decentralised and Privatised Irrigation System Management. In Deregulation, decentralisation and privatisation in irrigation: state functions move to the free market; DVWK Bulletin no 20; pp 188-216; Bonn 1997.
Hofwegen P.J.M. van, and E. Schultz (eds. 1997)
Financial Aspects of Water Management, Proceedings of the 3rd Netherlands National ICID Day , Balkema Rotterdam

IDB (1997)
> *Integrated Water Resources Management: Strategy Background Paper,* Draft October 24, 1997. IDB Washington

Lord., W.B. and M. Israel (1996)
> *A proposed strategy to encourage and facilitate improved water resources management in Latin America and the Caribbean.* IDB Washington,

Ostrom E. (1992)
> *Crafting institutions for self governing irrigation systems,* Centre for Self Governance, Institute for Contemporary Studies, San Fransisco.

Savenije H. (1997)
> *Concepts and Tools for Integrated Water Resources management,* IHE Delft

Small L. and M. Svendsen. (1990)
> *A framework for assessing irrigation performance;* Irrigation and Drainage Systems; Vol. 4/4, November 1990; Kluwer Academic Publishers, Dordrecht, the Netherlands.

Verhallen et.al. (1997) (in Dutch)
> *Integraal Waterbeheer;* Delft University of Technology.

World Bank (1993)
> Water Resources Management, a World Bank Policy Paper, Washington D.C.

ANNEX 1

Guidelines for Assessment of Institutional Framework - Constitutional Function

STAKEHOLDER: ....

	PRESENT SITUATION	"IDEAL" IWRM SITUATION	DESIRED SITUATION	GAP
A. STAKEHOLDERS aim: *to identify the stakeholder, its role and its relation with and perception of the role of other stakeholders in water resources management*				
1. What is the interest in water resources management of the institution you represent?		- for each agency their interests in water management are expressed in their mission statement. - for each user or beneficiary(group) interest in water use or control is expressed in production or protection statements.		
2. Does your institution carry out legislative, normative or planning functions?				
3. What water related problems in any of these functions have you encountered in exercising your task?		each stakeholder is able to indicate and formulate his water related problems directly or through his/her representative(s)		
4. What are problems and constraints within your organisation that you encountered in exercising your task?		each stakeholder is able to indicate and formulate organization related problems directly or through his/her representative(s)		
5. Are there organisations or groups that constrain effective implementation of your task?		each stakeholder is able to indicate which organizations constrain effective implementation of his/her task(s)		
6. Can you identify representatives or representations for these organisations?		each stakeholder is able to identify representatives of organizations constraining his/her task		
7. What organizations, platforms or individuals do you address to solve water related and organizational problems?		for all stakeholders there is a system of representation of interests that ultimately leads to a platform where all interests are weighed and decisions are made on use of water		
8. Are there constraining factors task that can not be addressed because there is nobody to address?		all water related aspects are within the jurisdiction of the coordination platform		

STAKEHOLDER:	PRESENT SITUATION	"IDEAL" IWRM SITUATION	DESIRED SITUATION	GAP
AWARENESS aim: *to identify the extent of understanding of the stakeholder of the need for IWRM*				
1. What is according your understanding IWRM?		all stakeholders are aware of the principles of IWRM and can explain that the basic feature is weighing of the different interest in use and control of water against social, economic, financial and environmental value and cost.		
2. Do you think that there is a need for IWRM? Why?		all stakeholders affected by other use and/or control activities have a need for IWRM to ensure their interest and not to harm others		
3. Do you think that all the other stakeholders and institutions related to water resources management know what IWRM is?		all stakeholders affected by other use and/or control activities know the concept of IWRM to ensure their interest and not to harm others		
4. Do you think that the IWRM approach should be on river basin (RB) level? What do you consider advantages and disadvantages?		IWRM is management of water systems preferably on RB level. advantages: 1. hydrological unity, 2. interdependence of different parts of the RB, 3. natural forum for conflict resolution, 4. obvious focus for data collection and analysis, 5 economy of scale, 6 opportunities for optimisation development and management etc. disadvantages 1. lack of coincidence with administrative and political boundaries, 2. cross cutting sectoral agencies, 3. scale either too large or too small for solving water problems, etc		
5. Are there problems or constraints of political, administrative or legal character that prevent water resources management on basin level?		In IWRM policies, legislation and administration are directed towards management on (sub)-basin level		

STAKEHOLDER:	PRESENT SITUATION	"IDEAL" IWRM SITUATION	DESIRED SITUATION	GAP
POLICY aim: *to identify the need for IWRM and the desired policy on IWRM*				
1. Is there a policy for IWRM? Can you describe the issues it addresses?		IWRM policies are directed at: - management at basin level, - participation of all stakeholders, - principle addressing interest-pay-say - principle addressing polluter pay - decentralised management - financial autonomy of RBA and 1st line agencies - set of standards and norms for use and control of water		
2. If such policy on IWRM does not yet exist, what elements have been developed and what has been done to implement them?				
3. Who is (or should be) responsible for the formulation of a national IWRM policy and who participates (or should participate) in this process?		IWRM policy is made in a collaborative effort by ministries responsible or involved in water use, water control and environment.		
4. Was the institute you represent involved in the process of (I)WRM policy development?		IWRM policies are made in a process of consultation of all stakeholders.		
5. What approaches have been, are being or should be used by the legislative, normative or planning institutions to make an IWRM policy				
6. Does the present (I) WRM policy address your needs as functionary adequately?		IWRM policies addresses the needs of all stakeholders but always will be a compromise between those needs		
7. What aspects of IWRM should be or are not yet dealt with?		IWRM deals with all aspects related to use and control of surface and ground water in qualitative and quantitative terms.		
8. How do you think IWRM should be organized?				

9. What do you consider the role of Government should be in the implementation of the IWRM policy?	Government takes care of the first order interests by providing adequate legislation, control and supervision on the implementation of these management activities.
10. Do you believe that the government should delegate functions in IWRM to the private sector? How would this delegation affect your functioning within you institution?	IWRM not necessarily requires full private sector participation nor full government responsibility. The effectiveness of PSP depends completely on the strength and capabilities of the private sector and the development policies of the government.
11. In what way do you think the private sector can play an effective role of in IWRM?	Private sector can participate on different levels and in different ways through: - service contracts for O&M of part of the assets - management contracts for O&M of all assets - lease contract for O&M of the assets and collection of fees from which the contractor makes its profits as well as pays its cost - concessions for O&M of the assets and new investments as and when required paid by collection of fees. - divestiture: the assets are owned and managed by a private company. Government may be (majority) shareholder to execute the care function

STAKEHOLDER:	PRESENT SITUATION	"IDEAL" IWRM SITUATION	DESIRED SITUATION	GAP
LEGAL FRAMEWORK aim: *to identify the present legal framework and its effectiveness and deficiencies from the perspective of the stakeholder and the changes under consideration.*				
1. Is there a law that regulates ownership, protection, distribution and use of water?		IWRM requires legislation on ownership and right to use or control water.		
2. Does the institute you represent feel protected or supported by the present legislation?		Legislation basically protects the rights to use or control water of all stakeholders but will provide priorities for use for the interest of society at large.		
3. Do you believe that the present legislation is really applicable on national level and responds to the needs of all the social, regional and local sectors having interest in water resources management?				
4.. Does present legislation provide you with sufficient tools to be applied in exercising your task effectively?		Legislation provides through regulations quantitative and qualitative standards, criteria and norms for each specific use, control and disposal of water.		
5. Does present legislation limit or constrain the execution of your task in an efficient and effective manner?		IWRM is based on legislation which allows efficient and effective use and control of water within the norms and standards set. This means a decentralised management by (financially) autonomous agencies with a great level of posteriori control and accountability.		
6. What aspects of legislation should be changed to allow you to exercise your task in an efficient, and effective manner?		- Legislation provides autonomy to stakeholders to use or control water within the set norms, standards and priorities for use for the interest of society at large. - legislation provides the framework of coordination and negotiation on water use, control and disposal - legislation provides mechanisms for accountability and conflict resolution		

7. Who do you consider should be responsible by law for resolving conflicts on water use?	IWRM provides a platform with decision power also related to resolution of conflicts between different interests.		
8. Do you think you should be consulted in case of legal reform?	In the preparation of IWRM legislation, stakeholders or their representatives are consulted.		
9. Is in your opinion law enforcement adequate to carry out your tasks effectively? What are the constraints and what should be improved?	Adequate and reliable law enforcement and transparent procedures is a prerequisite for effective IWRM.		
10. Is there legislation which allows private sector participation (PSP) in IWRM?			

STAKEHOLDER:	PRESENT SITUATION	"IDEAL" IWRM SITUATION	DESIRED SITUATION	GAP
INSTITUTIONAL FRAMEWORK *aim: to identify the present institutional framework, it's effectiveness and deficiencies from the perspective of the stakeholder*				
1. Who's interest does your institution represent?		Stakeholders at constitutional level represent society at large and societies interest in social-economic-environmental-security		
2. What decisions does your institution make regarding water use, protection, conservation, planning or distribution?		At constitutional level decisions are made on : - principles of ownership and right of use of water - assignment of functions to water systems - norms and standards for water use, control and disposal - missions, tasks and responsibilities of organizations with delegated water management tasks - principles on cost recovery for use, control and disposal of water on organizational and use level - participation criteria (and mechanisms) of stakeholders on organisational and use level - accountability mechanisms These decisions become law, government or ministerial regulation or decree.		
3. What other agencies or bodies does your institution consult with respect to water use protection, conservation, planning, distribution or application of norms and guidelines?		In IWRM organizations on organizational and use level are autonomous so no instructions are given to individuals or other than those related to the legal framework and the establishment or dissolving of these organisations.		
4. Is there coordination between constitutional level stakeholders involved in policy making and legislation?		IWRM requires a platform on constitutional level for consultation to obtain the necessary input and advise for policy making and development of legislation.		
5. Who are these other stakeholders?		All ministries with interest in use and control of water, administration, and finance, National representatives of stakeholder organisations.		

6. What is the status of the agreements reached at this platform.	advise to be considered by politicians in preparation of laws, regulations and decrees.
7. Are the interests of the different regional use and organizational stakeholders represented in the coordination and decision making processes?	In IWRM the interests of the different use level and organizational level stakeholders are represented through national representation in the advisory platform.

STAKEHOLDER:	PRESENT SITUATION	"IDEAL" IWRM SITUATION	DESIRED SITUATION	GAP
FINANCIAL ARRANGEMENTS aim: *to identify the arrangements of financing water use and control activities, the deficiencies and constraints*				
1. From what resources are your agencies activities financed?		-Government agencies are financed from govt budgets -representatives of stakeholders other than government are financed from contributions from their stakeholders		
2. Does your agency finance water use or control activities? or does it allocate subsidies, finances infrastructure, executes programmes or defines budgets? If yes from which sources?		In IWRM all organizations are in principle financially autonomous. However, financial support from constitutional level agencies can be expected in the form of subsidies or loans to: - implement the care function of government - to stimulate development - to implement programmes exceeding organizational or operation level agencies interest constitutional level financial operations are financed out of central govt budgets or loans from ESAs (local or international)		
3. Are the cost related to your financing of water related activities recovered fully or partially? Through what mechanism?				
4. Has your institution to deal with policies on tariffs for cost recovery on water use on national level?				

STAKEHOLDER:	PRESENT SITUATION	"IDEAL" IWRM SITUATION	DESIRED SITUATION	GAP
HUMAN RESOURCES DEVELOPMENT aim: *to identify present HRD strategies and policies to cope with the (I)WRM requirements.*				
1. Do you think that you have sufficient capable and motivated staff in your organization to execute your task in water resources management? What are the problem areas?				
2. Do you have in your organisation a HRD strategy to meet the demands on staffing capacity and capability? Do you know what the training needs of the personnel are?		Each organization has a HRD strategy		
3. If yes, what is the basis of this HRD strategy?		The HRD strategy is based on IWRM policies, organization's management strategy and operational needs and requirements.		
4. Does this strategy contain programmes on: - development of skills and knowledge - development of career opportunities - development of job incentives / proper salaries, bonuses				
5. Do you have funds for your HRD programme in your budget? are these adequate? what is the origin of these funds? are these incidental or structural funds?		HRD is financed out of the organizations own resources. Cooperation between educational institutes and IWRM stakeholders should lead to actualisation of curricula for public and private pre-job and in-job education.		
6. Are you involved in HRD or preparation of programmes for HRD for other regional and local stakeholders?				

7. Do you have adequate access to training and education centres to provide you with the necessary programmes? Is there an adequate information system on the activities of training centres?	Through a close cooperation between educational institutions and the water sector stakeholders adequate programmes are presented.	
8. What training programmes should be implemented on national level?	IWRM on national level requires education and training in technical, managerial and social knowledge and skills.	
9. Do you involve research centres in the development of your IWRM capacity?		

ANNEX 2

Guidelines for Assessment of Institutional Framework -
Organizational Function

STAKEHOLDER:	PRESENT SITUATION	"IDEAL" IWRM SITUATION	DESIRED SITUATION	GAP
STAKEHOLDERS aim: *to identify the stakeholder, its role and its relation with and perception of the role of other stakeholders in water resources management*				
1. What is the interest of the institution you represent in water resources management?		- for each agency their interests in water management are expressed in their mission statement. - for each user or beneficiary(group) interest in water use or control is expressed in production or protection statements.		
2. What type of functions does your institution carry out in water resources management: administrative, normative or legislative?				
3. What are problems related to management and administration of water in exercising your task?		each stakeholder is able to indicate and formulate his water related problems directly or through his/her representative(s)		
4. What are problems and constraints within your organisation that you encountered in exercising your task?		each stakeholder is able to indicate and formulate organization related problems directly or through his/her representative(s)		
5. Are there organisations or (groups of) individuals on national, regional or local level that constrain effective implementation of your task?		each stakeholder is able to indicate which organizations constrain effective implementation of his/her task(s)		
6. Can you identify representatives or representations for these organisations?		each stakeholder is able to identify representatives of organizations constraining his/her task		
7. What organizations, platforms or individuals do you address to solve water allocation, distribution or administration problems.		for all stakeholders there is a system of representation of interests that ultimately leads to a platform where all interests are weighed and decisions are made on allocation, distribution and administration of water		

WATER ALLOCATION
aim: *to identify present water allocation and distribution practices among the main stakeholders*

STAKEHOLDER:	PRESENT SITUATION	"IDEAL" IWRM SITUATION	DESIRED SITUATION	GAP
1. What is the process used for allocating water between different users and uses?		the process of water allocation is transparent and stakeholders on organizational and operational level can explain this process in main lines.		
2. What do you consider important to be changed in the present process?				
3. What is the present system of water rights? Are these water rights permanent, temporary, transferable?		water rights are clearly defined, water rights are temporary to enable society to appropriate these right if this is in the interest of society, water rights are transferable		
4. What do you consider important to be changed or developed in the system of water rights.				
5. What are the priorities in water allocation in case of shortages?		priorities are clearly defined and protect the interest of society		
6. What do you consider important to be changed or developed in the allocation priorities.				
7. Who participates in the decision making process on water allocation?		all stakeholders are involved in the development of rules and procedures for decision making. The interests of the stakeholders are taken into account and procedures for reclamation are established.		
8. What criteria are used in decision making on water allocations, control and sanctioning?		criteria used in the decision making process on water allocation are clearly formulated and given a legal status.		
9. What laws and regulations define the decision making process on water allocation.?		the decision making process is arranged in the bye laws of the platform for IWRM. These bye-laws are based on the relevant national and regional laws and regulations.		
10. What are the main problems you encounter in the water allocation process?				

STAKEHOLDER:	PRESENT SITUATION	"IDEAL" IWRM SITUATION	DESIRED SITUATION	GAP
AWARENESS aim: *to identify the extent of understanding the need for IWRM of the stakeholder*				
1. What is according your understanding IWRM?		all stakeholders are aware of the principles of IWRM and can explain that the basic feature is weighing of the different interest in use and control of water against social, economic, financial and environmental value and cost.		
2. Do you think that there is a need for IWRM? Why?		all stakeholders affected by other use or control activities have a need for IWRM to ensure their interest and not to harm others		
3. Do you think that all the other stakeholders and institutions related to water resources management know what IWRM is?		all stakeholders affected by other use or control activities know the concept of IWRM to ensure their interest and not to harm others		
4. Do you think that the IWRM approach should be on river basin (RB) level? What do you consider advantages and disadvantages?		IWRM is management of water systems preferably on RB level. advantages: 1. hydrological unity, 2. interdependence of different parts of the RB, 3. natural forum for conflict resolution, 4. obvious focus for data collection and analysis, 5 economy of scale, 6 opportunities for optimisation development and management etc. disadvantages 1. lack of coincidence with administrative and political boundaries, 2. cross cutting sectoral agencies, 3. scale either too large or too small for solving water problems.		
5. Are there political, administrative or legal problems or constraints on national or regional level that prevent water resources management on basin level?		In IWRM policies, legislation and administration are directed towards management on (sub)-basin level		
6. What actions have been or are undertaken on river basin level related to planning, distribution conservation and study of the (water) resources and what is actually implemented in your institution.				

STAKEHOLDER:	PRESENT SITUATION	"IDEAL" IWRM SITUATION	DESIRED SITUATION	GAP
POLICY aim: *to identify the need for IWRM and the desired policy on IWRM*				
1. Is there a policy for IWRM? Can you describe the issues it addresses?		IWRM policies are directed at: management at basin level, participation of all stakeholders, principle addressing interest-pay-say, principle addressing polluter pay, decentralised management, financial autonomy of RBA and 1st line agencies, set of standards and norms for use and control of water, private sector participation.		
2. If such policy on IWRM does not yet exist, what elements have been developed and what has been done to implement them?				
3. Who is (or should be) responsible for the formulation of a national IWRM policy and who participates (or should participate)?		IWRM policy is made in a collaborative effort by ministries responsible or involved in water use, water control and environment.		
4. Has the institute you represent been involved in the process of (I)WRM policy development?		IWRM policies are made in a process of consultation of all stakeholders.		
5. What approaches have been, are being or should be used by the legislative, normative or planning institutions to make an IWRM policy				
6. Does the present (I) WRM policy address your needs as functionary adequately?		IWRM policies addresses the needs of all stakeholders but always will be a compromise between those needs		
7. What aspects of IWRM are not yet dealt with?		IWRM deals with all aspects related to use and control of surface and ground water in qualitative and quantitative terms.		
8. How do you think IWRM should be organized?				

STAKEHOLDER:	PRESENT SITUATION	"IDEAL" IWRM SITUATION	DESIRED SITUATION	GAP
LEGAL FRAMEWORK aim: *to identify the present legal framework and its effectiveness and deficiencies from the perspective of the stakeholder.*				
1. Is there a law that regulates ownership, protection, distribution and use of water?		IWRM requires legislation on ownership, protection and right to use or control water.		
2. Does present legislation protect the interests of your institution regarding ownership, protection, distribution and use of water?		Legislation basically protects the rights to protect, use or control water of all stakeholders but will provide priorities for use for the interest of society at large.		
3. Do you believe that the present legislation has real applicability on national level and responds to the needs of all the social, regional and local sectors with interest in water resources management?				
3. Does present legislation provide you with criteria, standards and norms to be applied in exercising your task?		Legislation provides through regulations quantitative and qualitative standards, criteria and norms for each specific use, control and disposal of water.		
4. Where does present legislation limit or constrain the execution of your task in an efficient, effective and sustainable manner?		IWRM is based on legislation which allows efficient and effective use and control of water within the norms and standards set. This means a decentralised management by (financially) autonomous agencies with a great level of posteriori control and accountability.		
5. What should be changed in present legislation to allow you to exercise your task in an efficient, effective and sustainable manner?		- Legislation provides autonomy to stakeholders to use or control water within the set norms, standards and priorities for use for the interest of society at large. - legislation provides the framework of coordination and negotiation on water use, control and disposal - legislation provides mechanisms for accountability and conflict resolution		

6. Does the institution you represent establish norms or standards for water resources management? Are you presently involved in development of these? Which are these?			
7. What institution do you think should be responsible by law to resolve conflicts on water use?	IWRM provides a platform with decision power also related to resolution of conflicts between different interests.		
8. Are, were or will you be consulted in legal reform?	In the preparation of IWRM legislation, stakeholders or their representatives are consulted.		

STAKEHOLDER:	PRESENT SITUATION	"IDEAL" IWRM SITUATION	DESIRED SITUATION	GAP
INSTITUTIONAL FRAMEWORK aim: to identify the present institutional framework, it's effectiveness and deficiencies from the perspective of the stakeholder				
1. Who's interest does your institution represent?		Stakeholders at organizational level represent either: - general sector interests in social-economic-environmental-security, or - water sector interests (water supply, irrigation etc) or - government interest (care function)		
2. What decisions does your institution take regarding use, planning, protection, distribution and conservation of water?		At organizational level decisions are made on : - ownership and right or use of water - procedures for weighing interests and decision making - assignment of functions to water systems - allocation of water resources for use and control functions - procedures and priorities in case of scarcity or drought - local norms and standards for water use, control and disposal within the constitutional norms - mechanism for inspection, control and policing - mechanisms for cost recovery for use, control and disposal of water on organizational and use level - participation of stakeholders on organisational and use level - accountability mechanisms - investments in WRM infrastructure These decisions become local regulations or agreements (water contracts).		
3. Does your institution receive instructions from other agencies on application of norms and standards regarding planning, distribution, conservation or protection of water resources?		In IWRM organizations on organizational and use level are autonomous so no instructions are given other than those related to the legal framework and the establishment or dissolving of these organisations.		
4. Does your institution give instructions to other agencies on application of norms and standards regarding planning, distribution, conservation or protection of water resources?		IWRM requires a platform on organizational level for consultation to obtain the necessary input and advise for decision making on water use and control.		

5. Is there a coordination or consultation platform on regional or local level for the stakeholders with respect to decision making on planning, distribution, conservation or protection of water resources? What is the status of the agreements reached at this platform?	In IWRM all stakeholders are represented in a board which approves water allocation management plans. These plans are the basis for agreements between organizational functions and operational functions at regional and local level.			
6. Who are the other stakeholders in this coordination or consultation platform?	Stakeholders at organizational level are: - environmental protection agencies - economic sector interest (agriculture, industry, fisheries etc) - water sector interests (water supply, irrigation, power etc) and - national and local government(care function)			
7. Are the interests of all regional agencies and local water users represented in the coordination and decision making on planning, distribution, conservation or protection of water resources?				
8. How are the interests of your institution represented at national level in the development of policies, laws, norms and standards?				

STAKEHOLDER:	PRESENT SITUATION	"IDEAL" IWRM SITUATION	DESIRED SITUATION	GAP
FINANCIAL ARRANGEMENTS aim: to identify the arrangements to finance planning, coordination, distribution and policing activities, its effectiveness and deficiencies				
1. From where do the funds come that finance your institution's activities ?		An IWRM principle that can be applied is Interest-Pay-Say IWRM activities on organizational level are financed through contributions or charges from the different clients and beneficiaries of the services provided by the river basin authority based on actual cost. Activities of representatives of service and client organizations and other stakeholders are financed through contributions from those stakeholders.		
2. Does your agency finance water use activities? If yes through what mechanisms and from which sources?		The platform (RBA) finances activities for their own use and activities that exceed interests of specific water services. These activities are financed from contributions of stakeholders (which include national, regional and local government!)		
3. Are the cost related to your water use activities recovered fully or partially? (capital cost, recurrent cost, O&M cost etc.) If so, through what mechanism?		In the ultimate IWRM situation all costs for organizational level operations are fully recovered from the water services. Subsidies might be applied as means for development and for extra organizational interest. Also subsidies might be applied to exercise the care function of government.		
4. Is your institution involved in the setting of the tariffs related to recovery of costs of water use and water management on local and regional level? What is the basis and what are the criteria for tariff setting?		Tariffs for IWRM are set by the platform of stakeholders within tariff regulations imposed by government. Tariffs are based on real cost to provide the services. Tariffs of service delivery for specific use are set by the service provider in consultation with their clients within the tariff regulations imposed by government.		
5. Who collects the charges		Charges are collected directly from the client water services		

STAKEHOLDER:	PRESENT SITUATION	"IDEAL" IWRM SITUATION	DESIRED SITUATION	GAP
HUMAN RESOURCES DEVELOPMENT aim: *to identify present HRD strategies and practices and its effectiveness and deficiencies*				
1. Do you think that you have sufficient, capable and motivated staff in your organization to execute your task in water resources management? What are the problem areas?		Each organization has sufficient, capable and motivated staff.		
2. Do you have in your organisation a HRD strategy to meet the demands on staffing capacity and capability? Do you know what the training needs of the personnel are?		Each organization has a HRD strategy		
3. If yes, what is the basis of this HRD strategy?		The HRD strategy is based on IWRM policies, organization's management strategy and operational needs and requirements.		
4. Does this strategy contain programmes on: - development of technical skills and knowledge - development of social skills and knowledge - development of career opportunities - development of job incentives, salaries, bonuses				
5. Do you have funds for your HRD programme in your budget? are these adequate? what is the origin of these funds? are these incidental or structural funds?		HRD is financed out of the organizations own resources. Cooperation between educational institutes and IWRM stakeholders should lead to actualisation of curricula for public and private pre-job and in-job education.		
6. Are you involved in HRD or preparation of programmes for HRD for other regional and local stakeholders?				
7. Do you have agreements or adequate access to training and education centres to provide you with the necessary programmes? Is there an adequate information system on the activities of training centres?		Through a close cooperation between educational institutions and the water sector stakeholders adequate programmes are presented.		

ANNEX 3

Guidelines for Assessment of Institutional Framework -
Operational Function

STAKEHOLDER:	PRESENT SITUATION	"IDEAL" IWRM SITUATION	DESIRED SITUATION	GAP
A. STAKEHOLDERS aim: *to identify the stakeholder, it's role and it's relation with and perception of the role of other stakeholders in water use and water resources management*				
1. What is the interest in water or its management of the organization that you represent?		- for each agency (service providers) their interests in water and water management are expressed in their mission statement. - for each user or beneficiary(group) interest in water use or control is expressed in production or protection statements.		
2. What activities or functions related to water use and discharge of effluents does your organization carry out.		- each agency (service provider) can describe their function in acquisition, conveyance, distribution delivery and water supply, water level control, and effluent collection or discharge activities. - each stakeholder can describe its water use and effluent activities		
3. What technical, administrative, distribution, quantity and quality type of are water related problems have you encountered in exercising your task or in water use?		each stakeholder is able to indicate and formulate his water related problems directly or through his/her representative(s)		
4. As a service provider or as a water user, have you encountered organizational problems in exercising your task?		each stakeholder is able to indicate and formulate organization related problems directly or through his/her representative(s)		
5. What organisations or individuals constrain effective implementation of your task?		each stakeholder is able to indicate which organizations constrain effective implementation of his/her task(s)		
6. Can you identify representatives or representations for these organisations?		each stakeholder is able to identify representatives of organizations constraining his/her task		
7. What organizations, platforms or individuals do you address to solve water management and organizational problems?		for all stakeholders there is a system of representation of interests that ultimately leads to a platform where all interests are weighed and decisions are made on use of water		

STAKEHOLDER:	PRESENT SITUATION	"IDEAL" IWRM SITUATION	DESIRED SITUATION	GAP
WATER ALLOCATION aim: *to identify present water allocation and distribution practices related to particular stakeholders*				
1. What is the process used for allocating water between different users and uses?		the process of water allocation is transparent and stakeholders on organizational and operational level can explain this process in main lines.		
2. Do you participate in the decision making process on water allocation?		all stakeholders are involved on the development of rules and procedures for decision making within the margins provided by law. In these procedures the interests of the stakeholders are taken into account and procedures for reclamation are established.		
3. What do you consider important to be changed in the present process?				
4. What are your rights to use water? Are these water rights permanent, temporary, transferable?		- water rights are clearly defined - water rights are temporary to enable society to appropriate these right if this is in the interest of society - water rights are transferable		
5. What do you consider important to be changed or developed in the system of water rights.				
6. What are the priorities in water allocation in case of shortages?		priorities are clearly defined and protect the interest of society		
7. What do you consider important to be changed or developed in the system of allocation priorities.				

STAKEHOLDER:	PRESENT SITUATION	"IDEAL" IWRM SITUATION	DESIRED SITUATION	GAP
AWARENESS *aim: to identify the extent of understanding the need for IWRM of the stakeholder*				
1. What does according to you IWRM mean?		all stakeholders are aware of the principles of IWRM and can explain that the basic feature is weighing of the different interest in use and control of water against social, economic, financial and environmental value and cost.		
2. Do you think that there is a need for IWRM? Why?		all stakeholders affected by other use and/or control activities have a need for IWRM to ensure their interest and not to harm others		
3. Do you think that all the other organizations related to water management know what IWRM is about?		all stakeholders affected by other use and/or control activities know the concept of IWRM to ensure their interest and not to harm others		

STAKEHOLDER:	PRESENT SITUATION	"IDEAL" IWRM SITUATION	DESIRED SITUATION	GAP
POLICY aim: *to identify the desired policy on IWRM*				
1. Is there a policy for IWRM? Can you describe the issues it addresses?		IWRM policies are directed at: management at basin level, participation of all stakeholders, principle addressing interest-pay-say, principle addressing polluter pay, decentralised management, financial autonomy of RBA and 1st line agencies, set of standards and norms for use and control of water, private sector participation.		
2. If such policy on IWRM does not yet exist, what elements have been developed and what has been done to implement them?				
3. Who is (or should be) responsible for the formulation of a national IWRM policy and who participates (or should participate) in this process?		IWRM policy is made in a collaborative effort by ministries responsible or involved in water use, water control and environment.		
4. Has the institute you represent been involved in the process of (I)WRM policy development?		IWRM policies are made in a process of consultation of all stakeholders.		
5. What approaches have been, are being or should be used by the legislative, normative or planning institutions to make an IWRM policy				
6. Does the present (I) WRM policy address your needs as operator or user adequately?		IWRM policies addresses the needs of all stakeholders but always will be a compromise between those needs		

		IWRM deals with all aspects related to use and control of surface and ground water in qualitative and quantitative terms.	
7. What aspects of IWRM should be or are not yet dealt with?			
8. How do you think IWRM should be organized?			

STAKEHOLDER:	PRESENT SITUATION	"IDEAL" IWRM SITUATION	DESIRED SITUATION	GAP
LEGAL FRAMEWORK *aim: to identify the present legal framework and its effectiveness and deficiencies from the perspective of the stakeholder and the changes under consideration.*				
1. Is there a law that regulates ownership and use of water?		IWRM requires legislation on ownership and right to use or control water.		
2. Does the organisation that you represent feel that present legislation protects its interests? does it respond to the needs of all social, regional and local interests in managing water resources?		Legislation basically protects the rights to use or control water of all stakeholders but will provide priorities for use for the interest of society at large including all social, regional and local water related interests.		
3. Does present legislation provide you with appropriate criteria, standards and norms to be applied in exercising your task?		Legislation provides through regulations quantitative and qualitative standards, criteria and norms for each specific use, control and disposal of water.		
4. Does present legislation limit or constrain the execution of your task in an efficient, effective and sustainable manner?		IWRM is based on legislation which allows efficient and effective use and control of water within the norms and standards set. This means a decentralised management by (financially) autonomous agencies with a great level of posteriori control and accountability.		
5. What aspects should be changed in present legislation to allow you to exercise your task in an efficient, effective and sustainable manner?		- Legislation provides autonomy to stakeholders to use or control water within the set norms, standards and priorities for use for the interest of society at large. - legislation provides the framework of coordination and negotiation on water use, control and disposal - legislation provides mechanisms for accountability and conflict resolution		

Question	Description				
6. Can you describe the (accountability) mechanism that other institutions apply to control the activities of your organization related to delivery of water services, cost for service delivery and tariff setting?	Each service provider has an external and internal performance assessment procedure in its accountability mechanism which includes service delivery, use of resources for service delivery (financial, human, water, land etc.) Such mechanisms are oriented towards tutelage (usually government) and client relationships. Such mechanism is part of the regulations/ service agreements				
7. Can you describe the mechanism that you apply to control the activities related to water use and waste water of your clients (users)?	Each service provider has a Management Information System that provides information on water use and waste water of your clients (users). Regulations in agreements ensure that this information can be made available.				
8. Do you think that the law can be applied effectively and that law enforcement is adequately arranged? What rights and means does your organisation have to control and sanction violations?	Establishment of a legal framework includes the development of an effective law enforcement mechanism.				
9. What institution do you presently address in case of conflict on water use?	IWRM provides a platform with decision power also related to resolution of conflicts between different interests.				
10. What institution should according to you be assigned by law to resolve conflicts on water use?					
11. Do you think that you should be consulted in legal reform?	In the preparation of IWRM legislation, stakeholders are consulted or through their representatives.				

STAKEHOLDER:	PRESENT SITUATION	"IDEAL" IWRM SITUATION	DESIRED SITUATION	GAP
INSTITUTIONAL FRAMEWORK aim: *to identify the present institutional framework, it's effectiveness and deficiencies from the perspective of the stakeholder*				
1. Who's interest does your organisation represent?		At operational level organizations - provide services and have their client relations (irrigation, water supply, power, waste water treatment) : - represent client groups (WUA, etc) - represent public interest (govt. Agencies) - represent specific interests (pressure groups)		
2. What the mission of your organization?		In IWRM the mission of all stakeholders at operational level is clearly formulated. The missions are generally for: - (semi-)public services: provide highest level of service at least cost - private enterprise: maximise return on investment - government agency: protect (sectoral) interests of local society at large - pressure groups: protect specific interests of part of society		
3. What decisions does your organization make regarding water allocation, use, distribution and conservation or water control?		Service providers: allocation of water in their sector ,priorities for allocation, water ordering procedures, distribution procedures, sanctions, price setting, investments, operation and maintenance, Client organisations: do not decide on use and control but only advise, negotiate and represent the clients in the process of decision making of service providers. Government: set criteria and standards for use, control and disposal of water within those established by constitutional and organizational level agencies. Pressure groups do not take any decision but only try to influence decision making		

4. Does the organisation that you represent provide extension or information services to their clients regarding water use, planning, distribution, protection and conservation of the water source.	Each service provider or coordinating agency has a public information service to achieve and effective and sustainable use of water resources in general and their specific use of water in particular.		
5. From whom or what agencies or organizations do you get your instructions regarding water use, planning, distribution, protection and conservation of the water source?	In ideal IWRM situations - the water services are autonomous but they have to obey the service agreements with their clients and the rules and regulations as established by the government and the RBA (standards and norms). - the local government acts in accordance with the national and local legislation and within that framework with the RBA agreements (procedures) - client organizations work autonomous within their own set of by-laws and regulations.		
6. Is there a mechanism for coordination and concertation for decision making on water uses and distribution between the different users on local, regional and sectoral level?	In IWRM is a coordination and decision making platform on(sub) basin level for planning, use, distribution, control and conservation of water resources		
7. Who are other stakeholders on operational level?	The RBA: as planner and controller of water use Other water services: irrigation, domestic and industrial water supply, hydropower, flood control, fisheries, environmental protection Water users (clients), industries, agriculture, fisheries, (hydro) power companies, Government: care function (environment, social minority groups etc)		

8. What is the status of the agreements reached at this platform.		In IWRM the coordination platform is on RBA/organizational level. The agreements on this level are binding for all stakeholders. On local level the coordination that will take place is mainly bilateral between services and clients (service and service agreements), services and government (local regulations) or services and services (mutual interests). The agreements can be formal or informal but must be within the framework established by govt and RBA.		
9. How is the representation decided and by who?		Representatives are elected by the official members of the organization that represents the members interests. The procedures for election are laid down in the bye laws of the organization.		
10. Do you have a service agreement with your clients or service provider? What is the content?		Service agreements are only established for those services which can be provided to clearly identifiable individuals or groups. These services comprise irrigation, water supply ,wastewater treatment and in certain cases drainage. These service agreements consist of two parts: transactions and accountability mechanism.

No service agreements are made for public services like flood control and environmental protection. | | |
| 11. How is the agreement on service provision established? | | In IWRM the agreements are established based on consultations and negotiations by the service provider of their clients. | | |
| 12. What process will start if clients do not receive their agreed service? | | IWRM includes a transparent accountability mechanism and accountancy system so clients can be well informed on what happened and why services were not provided as agreed. | | |

STAKEHOLDER:	PRESENT SITUATION	"IDEAL" IWRM SITUATION	DESIRED SITUATION	GAP
FINANCIAL ARRANGEMENTS aim: *to identify the present way of financing water management operations and water services, its effectiveness and deficiencies*				
1. From which sources are the activities of your organisation financed?		Water services including investments are paid out of charges and service fees. Investments can be subsidised if they are of extra operational level interest.		
2. Does your agency finance activities related to water use? does it provide subsidies, finance infrastructure/equipment executes programmes or allocates and approves budgets?		The water service can contribute in financing activities - of their clients to stimulate better use of water - of other operational level or organizational level agencies if it serves mutual interests.		
3. Are the cost related to your water use activities recovered fully or partially? (capital cost, recurrent cost, O&M cost etc.) by which mechanism?		Ultimately, cost are fully recovered from charges and service fees from the clients. Subsidies might be received from govt. In exercising their care function.		
4. Who sets the tariffs how is this done and		Tariffs are based on real cost and are determined in a		

ANNEX 4

Guidelines for Assessment of Institutional Framework -
Evaluation Matrix

Constitutional Level

Requirement	Present Situation	Desired Situation	Gap	Proposed Intervention
a system that allows effective development and implementation of laws and regulations,				
a system that allows decision making based on interests of all stakeholders,				
a system that allows all stakeholders to participate in decision making,				
a system that provides quantitative and qualitative standards for use,				
a system that provides quantitative and qualitative standards for effluents,				
a system that allows effective control and sanctioning of violations,				
a system that allows implementing agencies to take the necessary steps to secure and conserve the resource,				
regulations on effective and transparent accountability mechanisms.				
sufficient capable people to meet the IWRM demands of policy making, adapting legislation and all other activities				
a system that allows private sector participation				

ORGANIZATIONAL LEVEL				
Requirement	Present Situation	Desired Situation	Gap	Intervention
a decision making capacity on river basin level which reflects the interests of different uses and users,				
a clear regulatory framework with norms and standards for decision making,				
a system that provides reliable information on the availability, use and quality of surface and ground water in the basin.				
a system that allows analysis of several options-scenarios for interventions in development and use of water at basin level,				
an effective and transparent accountability mechanism				
power to control and sanction violations				
sufficient capable people to meet the IWRM demands on planning and management, control and development.				

OPERATIONAL LEVEL				
Requirement	Present Situation	Desired Situation	Gap	Intervention
a system that allows effective control of (1st line) managing agencies (accountability mechanism) by users/clients and basin authorities.				
a system that ensures representation of clients interests at and by the managing agency				
a system that allows the managing agency to recover its cost.				
a system that allows the managing agency to negotiate with its clients on the level of service it provides and recovery of its associated cost,				
power to control and sanction violations				
sufficient capable people to meet the IWRM demands, planning, development and management of services provided				
a system that allows market incentives to make most economic use of water through participation of private sector.